David Fisher Day

Catalogue of the Niagara Falls

A Catalogue of the Flowering and Fern-Like Plants Growing Without Cultivation in

the Vicinity of the Falls of Niagara

David Fisher Day

Catalogue of the Niagara Falls
A Catalogue of the Flowering and Fern-Like Plants Growing Without Cultivation in the Vicinity of the Falls of Niagara

ISBN/EAN: 9783743442887

Manufactured in Europe, USA, Canada, Australia, Japa

Cover: Foto ©berggeist007 / pixelio.de

Manufactured and distributed by brebook publishing software
(www.brebook.com)

David Fisher Day

Catalogue of the Niagara Falls

INTRODUCTION.

In December, 1886, the writer, in answer to the request of the Commissioners of the State Reservation at Niagara, promised to prepare for their use a catalogue of the plants growing upon the reservation and its vicinity. He had already on hand the record of his observations made in the neighborhood of the Falls, during a period of more than twenty years. But he well knew that in order to give to the promised catalogue, such a degree of accuracy and completeness as would make it of value to botanists and the public, it would be necessary to revise and renew his observations in the field. To this task, he devoted such leisure as was at his command during the year 1887. The results are presented in this Catalogue. Still he does not doubt that further investigations, made in the vicinity of the Falls, will considerably increase the number of species here recorded. In the more difficult genera of the *Cyperaceæ* and *Gramineæ*, demanding always in a large degree the skill of the specialist, there must be omissions, more or less numerous and important. Yet it is probable that no species, really characteristic of the flora of Niagara, has been overlooked. To aid him in making the list complete, the writer has regarded it as his duty to consult, so far as was within his power, the observations made in the neighborhood of the Falls, by all other botanists. It is, however, a matter for great regret that references to the botany of the Falls, especially in the reports of the earlier explorers, have proved so few in number. It seems probable that PETER KALM, the friend and correspondent of the great LINNE, left some record of the botanical observations, which he made during his visit at Niagara, in the year 1750. But, the author has failed to find any mention of its publication, either in the Swedish tongue or in an English translation. If his journal still exists, its publication, at the present day, could not but be welcomed as an important contribution to the literature of American botany. It seems not unlikely that the species of *Hypericum* and *Lobelia*, which bear his name, were discovered by him near Table Rock.

It is to be doubted whether either the elder or the younger MICHAUX visited the neighborhood of the great cataract, and it is certain that the enterprising spirit of PURSH brought him no nearer than the site of the present city of Elmira. NUTTALL, who botanized near the Falls sometime previous to the year 1818, mentions but one plant, *Utricularia cornuta*, as found by him in their vicinity. TORREY doubtless visited the region — possibly was familiar with it—yet, in his Flora of the State of New York, published in the year 1843, of the 1,511 species of plants, which he described, only fifteen are attributed to Niagara, and none of these, upon his own authority. In the *Flora of North America*, of TORREY and GRAY, published in 1838-1842, Niagara is mentioned as a station only five times.

The labors of later botanists have been far more useful in the preparation of the list. The MS. journals of the HON. GEORGE W. CLINTON, while engaged in his arduous labors upon the botany of Buffalo and its vicinity, have proved of the greatest value; and the *"Flore Canadienne"* of ABBE PROVANCHER and the *"Catalogue of Canadian Plants"* of PROFESSOR MACOUN, the learned and indefatigable botanist of the Canadian Geological Survey, and the *"Canadian Filicineæ,"* the joint work of PROFESSOR MACOUN and DR. BURGESS, of London, Ontario, have been of important service.

Wherever use has been made of these or of other authorities due acknowledgment of the obligation appears in the list.

The geology of Niagara Falls, as related to the flora, demands at this place some brief attention. On either side of the river, at a distance not very constant, glacial clays appear. At Clifton they form a considerable elevation. With them, in places, also appear the usual gravel drift of the same period. These make up the soil of the adjacent country. But, as Hall and Lyell long ago pointed out, the evidence is complete that the river once stood at a very considerable height above its present rocky boundaries. At Chippewa, in Prospect park and at other places, the gravel deposits of the river, made whilst it was occupying its higher level, are still to be seen. But before the river formed its present bed in the gorge below the cataract, it cut through and carried away some portion of its former deposit, leaving, as it subsided, a terrace, on either side, still easily traced. This gravel deposit, once extending continuously across the present river, covered all of what is now Goat Island. It is characterized by a great abundance of freshwater shells, of the genera *Unio, Melania, Lymnœa*, etc., identical in species with those still inhabiting the river. The vegetation of the island is that then which might be expected to luxuriate upon a deep calcareous soil, enriched with an abundance of organic matter.

On either side of the river, following its course northerly, for a long distance, the same deposit is met with, alternating with patches from which it has been removed, and in which the underlying rock has but a scanty covering of soil. Near DeVaux College, and immediately above the Devil's Hole, near Lewiston, the rock is almost naked. Yet, even in such places, growing in the crevices of the rock or fringing the verge of the chasm, are to be found some of our most interesting plants. In some places, long since the chasm was excavated, the high rocky bank has given way, through the action of small streams of water, and perhaps by the operation of other causes, forming sheltered recesses of considerable extent, in which a rich humus has accumulated, supporting a dense growth of timber, and giving a congenial situation to some of our rarer plants. Among such places may be named the Devil's Hole, Foster's Flat, above Queenston, and the whirlpool wood, on the Canadian side.

The Niagara Escarpment, as it has been called by the geologists of the State of New York, known at Lewiston as the "Mountain" and in Canada as "Queenston Heights," presents some features entitled to notice in this place. At the point where the river makes its way through it, its height above Lake Ontario is 374 feet, and above Lake Erie, 32 feet. Though it presents a northerly exposure, yet among the plants, growing upon its talus and below, are a number which belong rather to the south and southwestward, and are much more abundant in Ohio than in Western New York. The fact may be explained by the higher annual temperature at Queenston and Lewiston and thence northward to Lake Ontario, than prevails at the Falls or immediately southward.

The flora of Goat Island presents few plants which may be called uncommon in Western New York. For the rarer plants, included in the Catalogue, other localities must be visited. But it is still true that Goat Island is very rich in the number of its species. Probably no tract of land in its vicinity, so restricted in area, can be found, exhibiting so large a number. Its vernal beauty is attributable, not merely to its variety of plants, conspicuous in flower, but also to the extraordinary abundance in which they are produced. Yet it seems likely that there was a time, probably not long ago, when other species of plants, of great beauty, were common upon the island, but which are not now to be found there. It is hardly possible that several orchidaceous plants and our three native lilies did not once embellish its woods and grassy places. Within a little while the Harebell, *(Campanula rotundifolia, L.)*, has gone and the Grass of Parnassus, *(Parnassia Caroliniana, L.)*, is fast going. This is undoubtedly due to careless flower-gatherers, who have plucked and pulled without stint or reason. The same fate awaits the *Sanguinaria*, the *Diclytras* and the *Trilliums*, which do so much to beautify the island, unless the wholesale spoliation is soon arrested.

The suggestion may here be made, with great propriety, that pains be taken to re-establish upon the island the attractive plants which it has lost. The effort would entail but little expense, its success would be entirely certain and to many the pleasure of a visit to the island would be greatly enhanced. It would surely be a step, and not an unimportant one, in restoring the island to the state in which nature left it.

Frequent reference has been made in the Catalogue to localities in the neighborhood of the Falls, but not within the boundaries of the Reservation. For this, however, there is a sufficient reason in the fact that if the Catalogue had been confined within the limits of the Reservation, it would have been of far less value to the many who are now interested in botanical science; and to them alone can such a work be of much interest or value. As already stated, some of the rarest plants of western New York and Ontario grow in the neighborhood of Niagara river, but not within the confines of the Reservation.

Thus, it may be mentioned that in the wooded grounds adjoining DeVaux College and belonging to that institution, there may be found *Rhus aromatica*, Ait., *Liatris cylindracea*, Michx., *Aster ptarmicoides*, T. & G., *Asclepias quadrifolia*, Jacq., and *Morus rubra*, L. Among its rocks, perhaps there may still be found a specimen of *Pellaea atropurpurea*, Link., *Camptosorus rhyzophyllus*, Link., and *Asplenium Trichomanes*, L. The Devil's Hole, now almost inaccessible, was once a paradise of ferns—*Phegopteris Dryopteris*, Fée, being its chief rarity. The plateau of rock, which overlooks the ravine, produces *Arabis hirsuta*, Scop., and *Selaginella rupestris*, Spreng., elsewhere in this region quite uncommon. Between the "Mountain" and Lewiston, the explorer will find *Ranunculus multifidus,* Pursh. *Xanthoxylum Americanum,* Mill. *Houstonia cærulea,* L. and *Asplenium ebeneum,* Ait., rare plants in western New York. Queenston Heights have yielded us *Anemonella thalictroides,* Spach. *Asimina triloba,* Dunal, *Lupinus perennis,* L., *Frasera Carolinensis,* Walt. and *Celtis occidentalis,* L., —species scarcely found elsewhere in our vicinity. Of Foster's Flat, above Queenston, it may be said that the spring seems to visit its rocky fastnesses some weeks earlier than the table-land above. Among its uncommon plants, it may be mentioned that PROFESSOR MACOUN and DR. BURGESS have detected *Aspidium Lonchites,* Swartz, and *Aspidium Boolii,* Tuckm., — ferns whose accustomed range is a hundred miles northward. The woods near the whirlpool, on the Canadian side, produce in abundance *Cerastium arvense,* L., *Arctostaphylos Uva-ursi,* Spreng., *Castilleia coccinea,* Spreng., and the only sassafras trees known in the neighborhood of the Falls. The low land, near Clifton, on the Canadian side, only a few inches higher than the river, affords such uncommon plants as *Gentiana serrata,* Gunner, and *Parnassia Caroliniana,* Michx. *Calamintha Nuttallii,* Benth., still grows on the damp rocks, near the border of the river, and *Gerardia purpurea,* L., and *Utricularia cornuta,* Michx., appear sparingly in the wet and oozy soil near by.

In the preparation of this list, the practice has been followed, now almost universal, of indicating introduced plants, (which it will be seen are a large number) by giving their names in small capitals. The later changes in the nomenclature of the species have also been adopted, adding however, (in parentheses) the names under which they were described in the last edition of Gray's Manual.

Of the 909 species of plants named in the Catalogue 758 are native and 151 foreign.

The following table exhibits, synoptically, the number of species and genera belonging to each natural family of plants mentioned in the Catalogue.

Names of families.	No. of genera.	No. of species
Ranunculaceæ	10	32
Magnoliaceæ	2	2
Anonaceæ	1	1
Menispermaceæ	1	1
Berberidaceæ	4	4
Nymphaceæ	2	2
Papaveraceæ	2	2
Fumariaceæ	2	3
Cruciferæ	12	25
Capparidaceæ	1	1
Violaceæ	2	8
Polygalaceæ	1	3
Caryophyllaceæ	7	12
Portulacaceæ	2	3
Hypericaceæ	1	7
Malvaceæ	4	6
Tiliaceæ	1	2
Linaceæ	1	2
Geraniaceæ	4	6
Rutaceæ	1	1
Simarubiaceæ	1	1
Ilicaceæ	2	2
Celastraceæ	2	4
Rhamnaceæ	1	1
Vitaceæ	2	4
Sapindaceæ	3	8
Anacardiaceæ	1	6
Leguminosæ	13	33
Rosaceæ	15	42
Saxifragaceæ	6	10
Crassulaceæ	2	4
Hamamlidaceæ,	1	1
Haloragæ	1	3
Lythraceæ	1	1
Onagraceæ	5	11
Ficoidæ .,	1	1
Umbelliferæ	15	19
Araliaceæ	1	3
Cornaceæ	2	7
Caprifoliaceæ	6	14
Rubiaceæ	4	13

Names of families.	No. of genera.	No. of species.
Valerianaceæ	1	2
Dipsaceæ	1	1
Compositæ	38	97
Lobeliaceæ	1	3
Campanulaceæ	2	5
Ericaceæ	8	14
Primulaceæ	3	6
Oleaceæ	2	4
Apocynaceæ	2	3
Asclepiadaceæ	2	7
Gentianaceæ	2	3
Polemoniaceæ	2	2
Hydrophyllaceæ	1	2
Borraginaceæ	6	11
Convolvulaceæ	3	7
Solanaceæ	5	7
Scrophulariaceæ	13	26
Orobanchaceæ	3	3
Lentibulaceæ	1	2
Acanthaceæ	1	1
Verbenaceæ	2	3
Labiatæ	18	27
Plantiginaceæ	1	4
Amarantaceæ	1	2
Chenopodiaceæ	2	7
Phytolaccaceæ	1	1
Polygonaceæ	2	17
Aristolochiaceæ	1	1
Piperaceæ	1	1
Lauraceæ	2	2
Thymelaceæ	2	2
Eleagnaceæ	1	1
Santalaceæ	1	1
Euphorbiaceæ	2	6
Ceratophyllaceæ	1	1
Urticaceæ	8	13
Platanaceæ	1	1
Juglandaceæ	2	6
Betulaceæ	2	4
Cupuliferæ	6	15
Salicaceæ	2	16

2

Names of families.	No. of genera.	No. of species.
Coniferæ	5	6
Hydrocharidaceæ	2	2
Orchidaceæ	4	7
Iridaceæ	2	3
Smilacaceæ	1	2
Liliaceæ	16	20
Pontederiaceo	2	2
Juncaceæ	2	9
Typhaceæ	2	4
Araceæ	4	4
Lemnaceæ	3	3
Alismaceæ	3	4
Naiadaceæ	3	18
Cyperaceæ	5	54
Graminæ	31	68
Equisetaceæ	1	6
Ophioglossaceæ	2	3
Filices	12	27
Selaginellaceæ	2	3
Hydropterides	2	2
Total	410	909

CATALOGUE.

RANUNCULACEÆ.

Clematis Virginiana, L........Clematis. **Virgin's Bower.**
On the Canadian side, near Clifton, and elsewhere.

Anemone cylindrica, Gray.......Anemone. **Wind Flower.**
Goat Island and elsewhere.

Anemone Virginiana, L.........Anemone. **Wind Flower.**
Goat Island and near DeVaux College.
var. **alba,** Wood.
Goat Island.

Anemone dichotoma, L. (*A. Pennsylvanica,* L.)
Anemone. **Wind Flower.**
Goat Island and elsewhere.

Anemone nemorosa, L..........Anemone. **Wind Flower.**
Goat Island.
var. **quinquefolia,** Gray.
With the last.

Anemone Hepatica, L. (*Hepatica triloba,* Chaix.)....**Liverwort.**
Goat Island and near Lewiston. Rather rare.

Anemone acutiloba, Lawson. *(Hepatica acutiloba,* DC.)*
Liverwort.
Goat Island and elsewhere. Less rare.

Anemonella thalictroides, Spach.
(*Thalictrum anemonoides,* Michx.)
Near Brock's Monument, Ontario.

Thalictrum dioicum, L..................... ......**Meadow Rue.**
Goat Island, Lewiston and elsewhere.

Thalictrum Cornuti, L.
Meadow Rue.
Shores of the river above the Falls.

Thalictrum purpurascens, L................**Meadow Rue.**
Near Clifton, Canada.
var. **ceriferum,** Austin.
" Near Drummondsville, Niagara Falls." Macoun.

Ranunculus aquatilis, L .... **Water Crowfoot.**
var. **trichophyllus,** Chaix.
In pools above the falls on the Canadian side.

Ranunculus multifidus, Pursh .... **Buttercup.**
In a pool not far from the river's bank, above Lewiston.

Ranunculus Flammula, L .... **Spearwort.**
var. **reptans,** Meyer.
At the water's edge on Strawberry and Grand Islands, and,
probably, in similar situations nearer the Falls.

Ranunculus abortivus, L.
Goat Island and elsewhere.

Ranunculus sceleratus, L .... **Cursed Crowfoot.**
Lewiston, and near Suspension Bridge.

Ranunculus recurvatus, Poir .... **Hooked Crowfoot.**
Goat Island.

Ranunculus Pennsylvanicus, L .... **Bristly Crowfoot.**
Clifton, Canada.

Ranunculus fascicularis, Muhl .... **Early Buttercup.**
Near DeVaux College.

Ranunculus septentrionalis, Poir. (*R. repens, L.*) .. **Buttercup.**
Clifton, Canada. The large variety of R. repens, L. Common in
western New York.

RANUNCULUS BULBOSUS, L .... BUTTERCUP.
Goat Island. Introduced.

RANUNCULUS ACRIS, L .... BUTTERCUP.
Goat Island and elsewhere.

Caltha palustris, L .... **Cowslips. Marsh Marygold**
Lewiston, Clifton, Ontario.

Aquilegia Canadensis, L .... **Wild Columbine**
Goat Island and along the rocky banks of the river in many
places.

Hydrastis Canadensis, L .... **Hydrastis**
Cayuga Island, and probably elsewhere nearer the falls.

Actæa spicata, L .... **Herb Christopher. Red Cohosh**
var. **rubra,** Ait.
Goat Island, near DeVaux College.

Actæa alba, Bigelow .... **White Cohosh**
Goat Island.

Cimicifuga racemosa, Nutt .... **Black Snake-root**
Probably occurs near Queenston, Ont.

MAGNOLIACEÆ.

Liriodendron Tulipifera, L ......**White wood. Tulip-tree**
Goat Island. Not common. One fine specimen is growing near
the carriage way on the north side of the island.

Magnolia acuminata, L ....**Cucumber tree**
"Near the falls of Niagara." *Provancher.* Not seen by us.

ANONACEÆ.

Asimina triloba, Dunal ....**Papaw**
Queenston Heights, Ontario, *Macoun.*

MENISPERMACEÆ.

Menispermum Canadense, L ....**Moon seed**
Common both in New York and Ontario, but not noticed by us
near the Falls. Doubtless overlooked.

BERBERIDACEÆ.

BERBERIS VULGARIS, L.................................BARBERRY.
Goat Island, near Lewiston.

Caulophyllum thalictroides, Michx ....**Blue Cohosh**
Goat Island. Abundant.

Jeffersonia diphylla, Pers...Twin Leaf. Rheumatism Root
Niagara Falls. *Clinton.* Very rare.

Podophyllum peltatum, L........May Apple. Mandrake
Goat Island. Abundant.

NYMPHACEÆ.

Nymphæa tuberosa, Paine.......... ....White Water Lily
Abundant in shallow places in the river, some distance above the
Falls.

Nuphar advena, Ait...................Yellow Water Lily
With the last.

PAPAVERACEÆ.

CHELIDONIUM MAJUS, L.................................CELANDINE.
Clifton, Ontario.

Sanguinaria Canadensis, L....................Blood Root
Goat Island, where it has been found producing pink flowers.

FUMARIACEÆ.

Diclytra Cucullaria, DC. (*Dicentra Cucullaria,* DC.)
Dutchman's Breeches
Goat Island. Abundant.

Diclytra Canadensis, DC. (*Dicentra Canadensis,* DC.)
Squirrel Corn
Goat Island. Abundant. Between the two species numerous hybrids have been noticed on Goat Island.

Corydalis glauca, Pursh . . . . .Corydalis
Occurs at Tonawanda, and should be found near the Falls.

CRUCIFERÆ.

ALYSSUM CALYCINUM, L. . . . . .ALYSSUM.
Near Brock's monument, Ontario.

DRABA VERNA, L. . . . . .WHITLOW GRASS.
Introduced on Goat Island, but perhaps not established.

Dentaria diphylla, Michx . . . . . .**Pepper Root. Crinkle Root**
Goat Island.

Dentaria laciniata, Muhl.
Goat Island.

Cardamine rhomboidea, DC . . . .**Spring Cress**
var. **purpurea,** Torr.
Goat Island and elsewhere. The typical form, probably, may be found in the low ground near Clifton, Ontario.

Cardamine hirsuta, L. . . . .**Bitter Cress**
Goat Island.

Arabis lyrata, L. . . . .**Rock Cress**
Goat Island. The Three Sisters. And along the gorge to Lewiston.

Arabis hirsuta, Scop.
Near Devaux College, and at Lewiston, and on the opposite side of the river.

Arabis lævigata, Poir.
Devil's Hole, Queenston Heights, Ontario. *Macoun.*

Arabis Canadensis, L. . . . .**Sickle Pod**
Goat Island, and elsewhere.

Arabis perfoliata, Lam. . . . .**Tower Mustard**
Near Clifton, Ontario.

Arabis Drummondii, Gray.
Lewiston. Not common.

Barbarea præcox, R. Br......................Early Winter Cress.
Brock's monument, Ontario. *Macoun.*

Barbarea vulgaris, R. Br.....................**Winter Cress**
Road sides near the falls.

Erysimum cheiranthoides, L.........**Worm-seed Mustard**
Margin of the river above the Falls.

Sysimbrium officinale, Scop.......................Hedge Mustard.
Road sides and waste places near the Falls.

Brassica Sinapisastrum, Boiss.............. .....Mustard. Charlock.
Abundant on both sides of the river.

Brassica nigra, Koch..............................Black Mustard.
Between the Falls and DeVaux College.

Nasturtium officinale, R. Br........................Water Cress.
Near the river's edge above the Falls. Clifton, Ontario.

Nasturtium palustre, DC.....................**Marsh Cress**
In damp places above the Falls.

Nasturtium lacustre, Gray.....................Lake Cress
In the river above the Falls.

Nasturtium Armoracia, Fries.......................Horse Radish
"At Niagara Falls." *Macoun.*

Capsella Bursa-pastoris, Mœnch.................Shepherd's Purse
Common everywhere.

Lepidium Virginicum, L...............**Wild Peppergrass**
Road sides, near the village.

Lepidium campestre, R. Br.....................Field Peppergrass
"Clifton, near Niagara Falls." *Macoun.*

CAPPARIDACEÆ.
Polanisia graveolens, Raf.
"Abundant on the sands at Niagara." *Macoun.*
Plentiful at the foot of Lake Erie.

CISTACEÆ.
Helianthemum Canadense, Michx.............**Frost Weed**
Common in dry places in Western New York, no doubt occurring
near DeVaux College, and at Lewiston.

Lechea major, Michx....................... ......**Pinweed**
Probably occurs with the last.

Lechea minor, Lam..............................**Pinweed**
Probably occurs with Helianthemum Canadense.

VIOLACEÆ.

Ionidium concolor, Benth. and Hook. (*Solea concolor,* Ging).
Green Violet
Goat Island. Foster's Flat, Ontario.

Viola blanda, Willd.................................**Violet**
Goat Island, and near Clifton, Ontario.

Viola palmata, L.................................**Violet**
var. **cucullata,** Gray. (*V. cucullata,* Ait.)
Goat Island and elsewhere.

Viola canina, L................................**Dog Violet**
var. **Muhlenbergii,** Gray. (*Viola canina* L., var. *sylvestris,*
Regel.)
Goat Island and elsewhere.

Viola rostrata, Muhl............ ....**Long-spurred Violet**
Goat Island.

Viola Canadensis, L.........................**Canada Violet**
Goat Island.

Viola pubescens, Ait.............. ....**Downy Yellow Violet**
var. **eriocarpa,** Nutt.
Goat Island.
var. **scabriuscula,** Torr. and Gray.
Goat Island.

POLYGALACEÆ.

Polygala verticillata, L.
Near DeVaux College. Queenston Heights, Ontario, *Macoun.*

Polygala Senega, L....................**Seneca Snake Root**
Both the narrow and the broad leaved varieties are to be found
near the whirlpool on both sides of the river.

Polygala incarnata, L.
Said by Douglass (1823) to have been found in rocky places on the
Niagara river near the Falls. *Macoun.*

CARYOPHYLLACEÆ.

DIANTHUS ARMERIA, L.............................DEPTFORD PINK.
Lewiston, scarce. In a field near Clifton, Ontario, plentiful.

SAPONARIA OFFICINALIS, L...............................SOAPWORT.
Goat Island and the mainland.

17

Silene stellata, Ait................ .... **Starry Campion**
Found by Douglass, in 1823, in dry, stony places on the Niagara
river. *Macoun.*

Silene antirrhina, L..................... **Sleepy Catch-fly**
Near DeVaux College and elsewhere.

SILENE NOCTIFLORA, L................. ......NIGHT-FLOWERING CATCH-FLY.
Above the Falls on the American side of the river.

LYCHNIS GITHAGO, Lam............................CORN COCKLE.
Fields on the main land.

CERASTIUM VISCOSUM, L..................... ..MOUSE-EAR CHICKWEED.
Goat Island and elsewhere.

CERASTIUM VULGATUM, L.....................MOUSE-EAR CHICKWEED.
Goat Island and elsewhere.

STELLARIA MEDIA, Smith...............................CHICKWEED.
Goat Island and elsewhere.

Stellaria longifolia, Muhl.......**Long-leaved Stitchwort.**
In damp, grassy places above the Falls.

ARENARIA SERPYLLIFOLIA, L..................... ....SAND WORT.
Road sides at Clifton and elsewhere.

Arenaria lateriflora, L.
Goat Island.
PORTULACACEÆ.

PORTULACA OLERACEA, L.............................. PURSLANE.
Waste places on the main land.

Claytonia Caroliniana, Michx...............**Spring Beauty**
Goat Island.

Claytonia Virginica, L.....................**Spring Beauty**
Goat Island.
HYPERICACEÆ.

Hypericum Kalmianum, L.......**Kalm's St. John's Wort**
Goat Island. "Rochers au bas de la chute de Niagara." *Provancher.* (*Fl. Canad. p.* 104.)

HYPERICUM PERFORATUM, L..................COMMON ST. JOHN'S WORT.
Goat Island and elsewhere.

Hypericum maculatum, Walt. (*H. corymbosum,* Muhl.)
Margin of the river above the falls.

Hypericum mutilum, L.
Wet places along the river above the falls.

Hypericum Canadense, L.
In similar places as the last.

3

Hypericum Ascyron, L. (*H. pyramidatum*, Ait.)

Great St. John's Wort

Grand Island. And probably nearer the falls.

Hypericum Virginicum, L. (*Elodes Virginica*, Nutt.)
In swampy places along the river above Clifton, Ontario.

MALVACEÆ.

ALTHÆA ROSEA, L.... ... HOLLYHOCK.
Clifton, Ontario. Escaped from cultivation.

MÁLVA ROTUNDIFOLIA, L.....Low MALLOW.
Goat Island and elsewhere.

MALVA SYLVESTRIS, L.....HIGH MALLOW.
Near the Devil's Hole.

MALVA MOSCHATA, L.....MUSK MALLOW.
Road sides on the main land.

ABUTILON AVICENNÆ, Gaert.....VELVET LEAF.
Waste places on the main land.

Hibiscus Moscheutos, L.....**Swamp Rose Mallow.**
Probably may be found along the river above the Falls, as it
grows in such situations near the foot of Lake Erie.

TILIACEÆ.

Tilia Americana, L.....**Linden. Basswood.**
Goat Island. An abundant and conspicuous element of its forest.

Tilia——? .....**Basswood**
Goat Island. The tree, here indicated, seems to be quite distinct
from the last. It may be readily distinguished by its bark,
which is as white as that of the white ash, *Fraxinus Americana, L.*

LINACEÆ.

Linum Virginianum, L .....**Wild Flax**
"Near Niagara Falls." *Macoun.*

LINUM USITATISSIMUM, L.....COMMON FLAX.
Occasionally seen on railroad embankments.

GERANIACEÆ.

Geranium maculatum, L.....**Wild Geranium**
Goat Island.

Geranium Robertianum, L .....**Herb Robert**
Goat Island.

Flœrkia proserpinacoides, Willd.
Goat Island.

Impatiens pallida, Nutt..........Impatience. **Wild Balsam**
Goat Island.

Impatiens biflora, Walt. (*I. fulva*, Nutt.)
Impatience. **Wild Balsam**
Goat Island and the main land.

Oxalis stricta, L.......................................Sorrel
Great Island.

RUTACEÆ.

Xanthoxylum Americanum, Mill.......... ....Prickly Ash
Near Lewiston.

SIMARUIBACÆ.

Ailanthus glandulosus, Desf..Ailanthus. **Tree of Heaven**
Spontaneous near Clifton, Ontario.

ILICACEÆ.

Ilex verticillata, Gray.......................Winter Berry
Chippewa. *Macoun.* Probably nearer the falls.

Nemopanthes Canadensis, DC.
Near Clifton, Ontario.

CELASTRACEÆ.

Celastrus scandens, L.............. ....Bittersweet
Goat Island.

Euonymus atropurpureus, Jacq..Burning Bush. **Wahoo**
Goat Island.

Euonymus Americanus, L ..Strawberry Bush. **Wahoo**
"Niagara." *Macoun.*
var. **obovatus,** Torr. and Gray.
"Hills around Niagara Falls." *Macoun,* on the authority of *Dr.
Maclagan.*

RHAMNACEÆ.

Ceanothus Americanus, L ..:   ...New Jersey Tea
Near DeVaux College.

VITACEÆ.

Vitis æstivalis, Michx......................Summer Grape
Goat Island and elsewhere.

Vitis riparia, Michx. (*V. cordifolia*, Michx.).......Frost Grape
Goat Island and elsewhere.

Vitis Labrusca, L........................**Wild Grape**
Erroneously attributed to the vicinity of the Falls, as a native, in
the " *Plants of Buffalo and Vicinity,*" (p. 26), and by *Provancher*
in the *Flore Canadienne*, (p. 112.) Occasionally spontaneous.

Ampelopsis quinquefolia, Michx........**Virginia Creeper**
Goat Island and elsewhere.

SAPINDACEÆ.

Staphylea trifolia, L.........................**Bladder Nut**
Foster's Flat, Ontario.

Æsculus glabra, Willd..........................**Buckeye**
Spontaneous near Lewiston. From the west.

Acer spicatum, Lam ....**Mountain Maple**
Goat Island, near the Horse-shoe falls.

Acer saccharinum, Wang.....................**Sugar Maple**
Goat Island. One of the most abundant trees.

Acer dasycarpum, Erhart.....................**White Maple**
Near Clifton, Ontario.

Acer rubrum, L................................**Red Maple**
Goat Island. A shade tree in the village.

Acer Negundo, L. *(Negundo aceroides,* Mœnch.*)* .....**Box Elder**
Planted in Prospect park, where it is now appearing spontaneously.

ACER PLATANOIDES, L............................NORWAY MAPLE
A common shade tree in the village.

ANACARDIACEÆ.

Rhus typhina, L.......................**Stag-horn Sumach**
Goat Island.

Rhus glabra, L..........................**Smooth Sumach**
Queenston Heights. *Macoun.*

Rhus venenata, DC.........................**Poison Sumach**
Swampy places above the Falls, near Clifton, Ontario.

Rhus Toxicodendron, L.....................**Poison Ivy**
Goat Island. Too plentiful.

var. **radicans,** Torrey.......................**Poison Ivy**
Goat Island.

Rhus aromatica, Ait....................**Aromatic Sumach**
Near DeVaux College and on the opposite side of the river.
Common in the places named.

LEGUMINOSÆ.

Lupinus perennis, L .... **Lupine**
Queenston Heights, Ontario.

MEDICAGO LUPULINA, L .... BLACK MEDICK
Above the Falls on the American side.

MELILOTUS OFFICINALIS, Willd .... YELLOW MELILOT
Above the Falls on the American side.

MELILOTUS ALBA, Lam ....SWEET CLOVER.
Goat Island and the main land.

TRIFOLIUM ARVENSE, L .... RABBIT'S-FOOT CLOVER.
Abundant near Lewiston.

TRIFOLIUM PRATENSE, L .... RED CLOVER.
Goat Island and elsewhere.

TRIFOLIUM REPENS, L .... WHITE CLOVER.
Goat Island and elsewhere.

TRIFOLIUM HYBRIDUM, L .... ALSIKE CLOVER.
American side of the river above the Falls. Lewiston.

TRIFOLIUM PROCUMBENS, L .... HOP CLOVER
Near Clifton, Ontario.

Robinia Pseudacacia, L .... **Common Locust**
A frequent shade-tree, often spontaneous.

Robinia viscosa, Vent ....**Clammy Locust**
Spontaneous in places near Lewiston.

Astragalus Cooperi, Gray .... **Cooper's Milk Vetch**
Goat Island, near the Horse-shoe fall.

Astragalus Canadensis, L .... **Milk Vetch**
Common on the islands in the river, near Tonawanda. It may be looked for nearer the Falls.

Desmodium nudiflorum, DC .... **Tick Trefoil**
Near DeVaux College.

Desmodium acuminatum, DC.
Near DeVaux College.

Desmodium pauciflorum, DC.
"Woods at Niagara Falls." *Macoun.*

Desmodium rotundifolium, DC.
Near DeVaux College.

Desmodium cuspidatum, Torr. and Gray.
Queenston Heights, Ontario. *Macoun,* on the authority of *Douglass.*

Desmodium Dillenii, Darlington.
Near DeVaux College. Queenston Heights, Ontario. *Macoun.*

Desmodium paniculatum, DC.
Near DeVaux College and on the opposite side of the river.

Desmodium Canadense, DC.
Near Clifton, Ontario.

Desmodium ciliare, DC.
Queenston Heights, Ontario. *Macoun.*

Lespedeza reticulata, Pers. *(L. violacea, Pers. var. sessiliflora, Torr. and Gray.)*
Near DeVaux College.

Lespedeza hirta, Ell.......................Bush Clover
"Queenston Heights and Niagara Falls," *Macoun.*

Lespedeza capitata, Michx.................Bush Clover
Near DeVaux College.

Vicia Cracca, L.............................Vetch. Tare
Near Clifton, Ontario.

Vicia Caroliniana, Walt........ Vetch
Goat Island.

Vicia Americana, Muhl.............................Vetch
Goat Island.

Lathyrus ochroleucus, Hook..................Wild Pea
Goat Island, and near the whirlpool, Ontario.

Lathyrus paluster, L.........................Marsh Pea
var. **myrtifolius,** Gray.
Goat Island and margins of the river above the falls.

Amphicarpæa monoica, Elliott................Hog Peanut
Goat Island.

Apios tuberosa, Mœnch..................Ground Nut
Doubtless near the falls, but not yet observed by us.

Gleditschia triacanthos, L...................Honey Locust
Spontaneous along the bank of the river near Lewiston.

ROSACEÆ.

Amygdalus Persica, L.....................................Peach
Spontaneous on Goat Island and near the Devil's Hole.

Prunus Americana, Marshall ....Wild Plum
Goat Island and elsewhere.

Prunus Cerasus, L........................ ...Common Red Cherry
Spontaneous on Goat Island, and abundant along roadsides below
the Falls.

Prunus Virginiana, L.... ....Choke Cherry
Near DeVaux College.

Prunus serotina, L..........................Black Cherry
Near Clifton, Ontario.

Spiræa salicifolia, L...**Common Meadow Sweet**. Spiræa
Above the Falls on the Canadian side.

Neillia opulifolia, Benth. and Hook. (*Spiræa opulifolia*, L.)
Nine-bark
Goat Island. The Three Sisters.

Rubus odoratus, L........**Purple Flowering Raspberry**
Goat Island and elsewhere.

Rubus triflorus, Rich.
Wet places near Clifton, Ontario.

Rubus strigosus, Michx........ ....**Red Raspberry**
Goat Island.

Rubus occidentalis, L..................**Black Raspberry**
Goat Island.

Rubus villosus, Ait..........................**Blackberry**
Goat Island.

Rubus Canadensis, L.......**Dewberry. Low Blackberry**
Goat Island. Lewiston.

Geum album, Gmelin ....**Avens**
Goat Island.

Geum Virginianum, L.
Goat Island. Chippewa. *Macoun*, on the authority of *Dr. Maclagan*.

Geum strictum, Ait.
Goat Island.

Geum rivale, L...............................**Purple Avens**
Wet places above Clifton, Ontario.

Waldsteinia fragarioides, Tratt........**Barren Strawberry**
Goat Island.

Fragaria Virginiana, Duchesne..........**Wild Strawberry**
Goat Island and the main land.

Fragaria vesca, L.......................**Wood Strawberry**
Goat Island.

Potentilla Norvegica, L.
Goat Island and the main land.

Potentilla Canadensis, L ....Cinquefoil
Goat Island.
var. **simplex,** Torr. and Gray.
Near DeVaux College.

Potentilla argentea, L .... Silvery Cinquefoil
Near DeVaux College. Clifton, Ontario.

Potentilla Anserina, L ....Silver Weed
Low grounds above the falls on both sides of the river.

Potentilla palustris, Scop ....Marsh Five-finger
Chippewa. Clinton.

Potentilla pilosa, Willd.
Near Clifton, Ontario. This is the plant called by us, *P. recta,*
L., in the "*Plants of Buffalo and Vicinity.*" The present deter-
mination was by *Prof. Macoun.*

Agrimonia Eupatoria, L ....Agrimony
Goat Island.

Rosa blanda, Ait ....Early Wild Rose
Goat Island.

Rosa Carolina, L ....Swamp Rose
Wet grounds, near Clifton, Ontario.

Rosa parviflora, Ehrt. (*R. lucida,* Ehrt.) ....Shining Rose
Goat Island.

Rosa rubiginosa, L ....Sweet Briar
Goat Island. Devil's Hole. Lewiston.

Rosa micrantha, Smith..............................Sweet Brier
Goat Island.

Pyrus Malus, L...Apple
Spontaneous on Goat Island and near the Devil's Hole and
Lewiston.

Pyrus communis, L................. ....Pear
Spontaneous on Goat Island.

Pyrus coronaria, L........................Wild Crab-apple
Near DeVaux College. Lewiston. Queenston. Queenston Heights,
Ontario.

Pyrus Aucuparia, Gaertn ....Rowan Tree. Mountain Ash
Within the gorge of the River, on the Canadian side, near the
Cantilever Bridge.

Cratægus coccinea, L ....Thorn
Goat Island.

Cratægus tomentosa, L...................... Black Thorn
Goat Island.

Cratægus Crus-galli, L................ .. Cockspur Thorn
Goat Island. Not common in Western New York; but here,
quite abundant.

Amelanchier Canadensis, Torr. and Gray....... Shad-bush
var. **Botryapium, Gray.**
Goat Island.
var. **oblongifolia, Gray.**
Goat Island. Clifton, Ontario.

SAXIFRAGACEÆ.

Saxifraga Virginiensis Michx...........Spring Saxifrage
Goat Island. Near De Vaux College. Lewiston.

Saxifraga Pennsylvanica, L....... .....Swamp Saxifrage
Cayuga Island, near LaSalle. Clinton.

Tiarella cordifolia, L...........................Mitre Wort
Goat Island.

Mitella diphylla, L...... ....Mitre Wort
Goat Island.

Mitella nuda, L......................Naked Mitre Wort
Near Chippewa. Clinton.

Chrysosplenium Americanum, Schw.....Golden Saxifrage
Wet grounds near Clifton, Ontario.

Parnassia Caroliniana, Michx.
Goat Island near the Horse-shoe Fall. Wet grounds near Clifton,
Ontario. Near the water's edge at the Whirlpool on the Canadian
side.

Ribes Cynosbati, L.......................Wild Gooseberry
Goat Island.

Ribes oxyacanthoides, L. (*R. hirtellum*, Michx.).Swamp Gooseberry. Wet places near Clifton, Ontario.

Ribes floridum, L.. ....Wild Black Currant
Along the descent to the Ferry on the Canadian side.

4

CRASSULACEÆ.

Penthorum sedoides, L................. Ditch Stone Crop
Damp places near Clifton, Ontario.

SEDUM ACRE, L..STONE CROP
Goat Island and the mainland. Abundant.

Sedum ternatum, Michx..........................STONE CROP
Attributed to rocky places on Niagara River by Douglass. Not seen by us.

SEDUM TELEPHIUM, L..............................LIVE-FOR-EVER
Near De Vaux College.

HAMAMELACEÆ.

Hamamelis Virginica, L ... . Witch Hazel
Near De Vaux College.

HALORAGEÆ.

Myriophyllum spicatum, L................. Water Milfoil
Niagara River above the Falls in shallow and quiet places.

Myriophyllum verticillatum, L............. Water Milfoil
With the last.

Myriophyllum heterophyllum, Michx....... Water Milfoil
Pools near Clifton, Ontario.

LYTHRACEÆ.

Nesæa verticillata, H. B. K............. Swamp Loosestrife
Found in several places along the shores of Niagara River near Lake Erie. Therefore very likely to occur near the Falls, but not yet observed by us.

ONAGRACEÆ.

Epilobium spicatum, L. (*E. augustifolium,* L.).... Willow Herb
Near De Vaux College.

EPILOBIUM HIRSUTUM, L.
Introduced near Clifton, Ontario. Perhaps not established.

Epilobium palustre, L.
var. **lineare,** Gray.
Near Clifton, Ontario.

Epilobium molle, Torr.
In wet places near Clifton, Ontario.

Epilobium coloratum, Muhl.
Above the Falls on the American side.

Ludwigia palustris, Ell.
Not yet seen by us near the Falls, but may be confidently looked
for.

Œnothera biennis, L...................**Evening Primrose**
Goat Island and elsewhere.

Œnothera pumila, L. (Œ. chrysantha, Michx.)..**Dwarf Evening
Primrose.**
Near the Cantilver Bridge on the Canadian side of the River.
Queenston Heights, Ontario. *Macoun.*

Gaura biennis, L.....................................**Gaura**
Near the Devil's Hole.

Circæa Lutetiana, L ....**Enchanter's Nightshade**
Goat Island.

Circæa alpina, L..... ....**Enchanter's Nightshade**
Damp and shady woods near Clifton, Ontario.

CUCURBITACEÆ.

Echinocystis lobata, Torr. and Gray........**Wild Cucumber**

Sicyos augulatus, L ....**Star Cucumber**
These two members of the Gourd family, occurring not rarely in
Ontario and Western New York, have not been observed by us
near the Falls. They may be expected.

FICOIDEÆ.

Mollugo verticillata, L.....................**Carpet Weed**
"On the railway track between Niagara Falls and Queenston."
Macoun.

UMBELLIFERÆ.

Hydrocotyle Americana, L.................**Penny Wort**
Damp and shady places near Clifton, Ontario.

Sanicula Canadensis, L...........................**Sanicle**
Goat Island.

Sanicula Marilandica, L ....**Sanicle**
Goat Island.

CONIUM MACULATUM, L............................POISON HEMLOCK
Road sides on the mainland.

Cicuta maculata, L.............. ...   .**Water Hemlock**
  Goat Island.

Cicuta bulbifera, L......................**Water Hemlock**
  Wet places near Clifton, Ontario.

CARUM CARUI, Koch.....................................CARAWAY
  Goat Island.

Sium cicutæfolium, Gmel., *(S. lineare,* Michx.)..**Water Parsnip**
  Wet places near Clifton, Ontario.

Pimpinella integerrima, Beuth. and Hooker.   (*Zizia integerrima,*
  DC.)
  Near De Vaux College.

Cryptotaenia Canadensis, DC..................**Hone Wort**
  Goat Island.

Osmorrhiza longistylis, DC...........**Smooth Sweet Cicely**
  Goat Island.

Osmorrhiza brevistylis, DC...........**Hairy Sweet Cicely**
  Goat Island.

Thaspium barbinode, Nutt.
  Near De Vaux College.   Foster Flat, Ontario.   *Macoun.*

Thaspium aureum, Nutt...............**Golden Alexanders**
  Goat Island.

Archangelica atropurpurea, Hoffm...........Great Angelica
  Wet grounds near Clifton, Ontario.

PEUCEDANEUM SATIVUM, Beuth. and Hook. (*Pastinaca sativa,* L.)..PARSNIP
  Along the descent to the Ferry on the Canadian side.

Heracleum lanatum, Michx...................**Cow Parsnip**
  Goat Island.   More abundant on the mainland.

DAUCUS CAROTA, L..CARROT
  Near the Cantilever Bridge on the Canadian side.

Erigenia bulbosa, Nutt...............**Harbinger of Spring**
  Goat Island.   Introduced.   Established ?

ARALIACEÆ.

Aralia nudicaulis, L...................**Wild Sarsaparilla**
  Goat Island.   The Three Sisters.

Aralia quinquefolia, Gray......................**Ginseng**
  Goat Island, but rare.

Aralia trifolia, Gray.......................**Dwarf Ginseng**
  Goat Island.

CORNACEÆ.

Cornus florida, L..............Flowering Dogwood
Goat Island. Near De Vaux College.

Cornus circinata, L'Her..........Round-leaved Dogwood
Goat Island near the Horse-shoe Fall.

Cornus sericea, L.................Silky Dogwood
Goat Island.

Cornus stolonifera, Michx..........Red Osier
Goat Island.

Cornus paniculata, L'Her..........Dogwood
Goat Island. Near De Vaux College.

Cornus alternifolia, L.
Not uncommon in Western New York. Probably overlooked.

Nyssa multiflora, Wang.............Pepperidge. Tupelo
Not a rare tree in Erie and Niagara Counties. Ought to be found near the Falls.

CAPRIFOLIACÆ.

Sambucus racemosa, L.............Red-berried Elder
var. **pubens, Watson.**
Goat Island.

Sambucus Canadensis, L..............Elder
Goat Island.

Viburnum Opulus, L....Snowball. Guelder Rose. High Cranberry.
Grand Island. Likely to be found on either side of the River near the Falls.

Viburnum acerifolium, L..............Arrow wood
Near De Vaux College.

Viburnum pubescens, Pursh..........Arrow wood
Goat Island. Near De Vaux College. Along the descent to the Ferry on the Canadian side.

Viburnum dentatum, L..............Arrow wood
Chippewa, Ontario. *Macoun*, on the authority of *Dr. Maclagan.* Common on Grand Island.

Viburnum nudum, L..............Withe rod
Probably in the wet grounds near Clifton, Ontario.

Viburnum Lentago, L....Sheep-berry. Sweet Viburnum
Wet grounds near Clifton, Ontario.

Triosteum perfoliatum, L.................**Horse Gentian**
Near De Vaux College.

Symphoricarpus racemosus, Michx.............**Snowberry**
var. **pauciflorus,** Robbins.
Goat Island. Near De Vaux College and elsewhere.

Lonicera ciliata, Muhl...................**Fly Honeysuckle**
Goat Island. Near Clifton, Ontario.

LONICERA TARTARICA, L......................TARTARIAN HONEYSUCKLE
Goat Island. Near De Vaux College. Well established.

Lonicera glauca, Hill. (*L. parviflora,* Lam., var.*Douglasii,* Gray.)
Goat Island.

Diervilla trifida, Moench...............**Bush Honeysuckle**
Near Clifton, Ontario.

RUBIACEÆ.

Houstonia cærulea, L.................**Bluets. Innocence**
Near Lewiston.

Houstonia purpurea, L.
var. **ciliolata,** Gray.
Goat Island. Near the Whirlpool, on both sides of the River.

Cephalanthus occidentalis, L.................**Button Bush**
Wet places near Clifton, Ontario.

Mitchella repens, L.....................**Partridge Berry**
Goat Island.

Galium Aparine, L..............................**Cleavers**
Goat Island and elsewhere.

GALIUM MOLLUGO, L.....................................BED STRAW
Goat Island. Introduced.

Galium pilosum, Ait...........................**Bedstraw**
Near De Vaux College. Queenston, Ontario. *Macoun.*

Galium circæzans, Michx.
Lewiston. Queenston Heights, Ontario. *Macoun.*

Galium lanceolatum, Michx.
Near De Vaux College. Near the Whirlpool. Ontario. *Macoun,*
on the authority of *Dr. Maclagan.*

Galium boreale, L ....**Northern Bed straw**
Goat Island. Near De Vaux College.

Galium trifidum, L.
var. **tinctorum,** Gray.
Goat Island

Galium asprellum, Michx . . . . **Rough Bedstraw**
Goat Island.

Galium triflorum, Michx . . . . **Sweet Bedstraw**
Goat Island. Woods near De Vaux College.

VALERIANACEÆ.

VALERIANA OFFICINALIS, L . . . . MILLEFLEUR. VALERIAN
Near the Cantilever Bridge on the Canadian side.

Valeriana dioica, L., var. **sylvatica,** Watson. (V. *sylvatica*. Rich.)
"Meadows, Niagara Falls." [Ontario.] *Macoun*, on the authority
of *Dr. Maclagan.* Not seen by us.

DIPSACEÆ.

DIPSACUS SYLVESTRIS, Huds . . . . TEASEL
Above the Falls, on the American side of the River. Near the
Devil's Hole.

COMPOSITÆ.

Eupatorium purpureum, L . . . . **Purple Thoroughwort**
Damp grounds near Clifton, Ontario.

Eupatorium perfoliatum, L . . . . . . **Boneset. Thoroughwort**
Above the Falls on the American side.

Eupatorium ageratoides, L . . . . **White Snakeroot**
Goat Island and elsewhere.

Liatris cylindracea, Michx . . . . **Button Snakeroot**
Near De Vaux College.

Solidago cæsia, L . . . . **Golden Rod**
Near De Vaux College.

Solidago latifolia, L . . . . **Golden Rod**
Goat Island.

Solidago bicolor, L . . . . **Golden Rod**
Goat Island. Near De Vaux College.

var. **concolor,** Torr and Gray.
Near De Vaux College.

Solidago ulmifolia, Muhl . . . . **Golden Rod**
Goat Island.

Solidago neglecta, Torr. and Gray . . . . **Golden Rod**
"Niagara Falls." *Macoun,* on the authority of *Dr. Burgess.*

Solidago arguta, Ait . . . . **Golden Rod**
Chippewa. *Macoun,* on the authority of *Dr. Maclagan.*

Solidago juncea, Hook. (*S. arguta,* Ait.) . . . . Golden Rod
Goat Island.

Solidago serotina, Ait . . . . Golden Rod
Goat Island. "Niagara District." *Macoun.*
var. **gigantea,** Gray.
Goat Island.

Solidago Canadensis, L . . . . Golden Rod
Goat Island. Near Clifton, Ontario.

Solidago nemoralis, Ait . . . . Golden Rod
Goat Island. Near Clifton, Ontario.

Solidago rigida, L . . . . Golden Rod
Near De Vaux College.

Solidago lanceolata, Ait . . . . Golden Rod
Margin of Niagara River above the Falls on the American side.

BELLIS PERENNIS, L . . . . ENGLISH OR TRUE DAISY
In a lawn at Clifton, Ontario, where it has maintained itself a
number of years.

Aster corymbosus, Ait . . . . Aster
Goat Island. Lewiston.

Aster macrophyllus, L . . . . Aster
Goat Island. Near De Vaux College.

Aster Novæ-Angliæ, L . . . . New England Aster
Goat Island and above the Falls on either side.

Aster patens, Ait . . . . Aster
Near De Vaux College,
var. **phlogifolius,** Nees.
With the typical variety.

Aster azureus, Lindl . . . . Aster
Near De Vaux College. La Salle. *Clinton.*

Aster cordifolius, L . . . . Aster
Near De Vaux College.

Aster sagittifolius, Willd . . . . Aster
Goat Island. "Niagara." *Macoun,* on the authority of *Dr.*
Maclagan.

Aster lævis, L . . . . Aster
Near De Vaux College.

Aster ericoides, L . . . . Aster
Near De Vaux College.

Aster multiflorus, L . . . . Aster
Goat Island.

Aster vimineus, Lam. (*A. puniceus*, L., var. *vimineus.* Gray.
Aster.
Near Clifton, Ontario.

Aster diffusus, Ait. (*A. miser*, Nutt.) .... Aster
Goat Island. Near Clifton, Ontario.

Aster Tradescanti, L .... Aster
Goat Island.

Aster paniculatus, Lam .... Aster
Goat Island. Near De Vaux College.

Aster prenanthoides, Muhl .... Aster
Near Clifton, Ontario.

Aster puniceus, L .... Aster
Goat Island.

Aster umbellatus, Mill. (*Diplopappus umbellatus*, Torr. and Gray.)
Aster.
Goat Island. Near Clifton, Ontario.

Aster ptarmicoides, Torr. and Gray .... Aster
Near De Vaux College. A species not common in our region, but
here rather abundant.

Erigeron bellidifolius, Muhl .... Poor Robin's Plantain
Goat Island.

Erigeron Philadelphicus, L .... Pink Fleabane
Near Clifton, Ontario.

Erigeron annuus, Pers .... Fleabane
Waste places on the mainland.

Erigeron strigosus, Muhl .... Fleabane
Goat Island.

Erigeron Canadensis, L .... Horseweed
Mainland above the Falls on the American side.

Antennaria plantiginifolia, Hook .... Everlasting
Lewiston.

Anaphalis margaritacea, Benth. and Hook. (*Antennaria margaritacea*, R. Br.) .... Pearly Everlasting
Near De Vaux College.

Gnaphalium polycephalum, Michx .... Everlasting
Near De Vaux College.

Gnaphalium uliginosum, L.
Damp places along road sides on the main land.

INULA HELENIUM, L. .... ELECAMPANE
Goat Island.

Polymnia Canadensis, L .... **Leaf-cup**
Along the descent to the Ferry on the Canadian side of the River, and at the Whirlpool and Foster's Flat, Ontario.

Silphium trifoliatum, L.
'Attributed to the Falls, by *Torrey*, on the authority of *Dr. Eddy.* Not observed by us.

Ambrosia trifida, L ... . **Rag Weed**
Margin of the River above the Falls.

Ambrosia artemisiæfolia, L .... **Rag Weed**
Goat Island and with the foregoing species.

Xanthium Canadense, Mill .... **Cockle-bur**
Above the Falls on the American side.
var. **echinatum, Gray.**
Chippewa, Ontario. *Macoun,* on the authority of Dr. Maclagan.

Heliopsis lævis, Pers .... **False Sun-flower**
Near Clifton, Ontario.

Rudbeckia hirta, L .... **Yellow Daisy**
Near Lewiston.

Rudbeckia laciniata, L .... **Cone Flower**
Wet places near Clifton, Ontario.

HELIANTHUS ANNUUS, L. .... COMMON SUN-FLOWER
Escaped near Lewiston.

Helianthus divaricatus, L .... **Sun-flower**
Goat Island.

Helianthus strumosus, L ....'....**Sun-flower**
Goat Island. Near DeVaux College.

Helianthus decapetalus, L .... **Sun-flower**
Damp places near Clifton, Ontario.

Bidens frondosa, L .... **Cockle.   Beggar's ticks**
Waste places on the main land.

Bidens connata, Muhl .... **Beggar's ticks**
Goat Island.

Bidens cernua, L.
Wet places near Clifton, Ontario.*

Bidens chrysanthemoides, Michx.
With the last and on the American side of the River above the Falls.

Bidens Beckii, Torr.
Chippewa, Ontario. *Mawun*, on the authority of *Dr. Maclagan.*
Niagara River near Grand Island.

Helenium autumnale, L........ .... **Sneeze-weed**
Goat Island and elsewhere in wet ground.

ANTHEMIS COTULA, L. (*Maruta Cotula, D. C.*)............. MAYWEED
Road sides on the main land.

Achillæa Millefolium, L........................... **Yarrow**
Goat Island.

CHRYSANTHEMUM LEUCANTHEMUM, L..................... OX-EYE DAISY
Goat Island and elsewhere.

CHRYSANTHEMUM PARTHENIUM, Pers....................... FEVERFEW
Lewiston. A garden scape.

TANACETUM VULGARE, L............................... .... TANSY
Near the Cantilever Bridge on the Canadian side.

ARTEMISIA VULGARIS, L............................... MUGWORT
Along the descent to the old landing on the American side of the
River, of the steamer "Maid of the Mist."

TUSSILAGO FARFARA, L ....COLTSFOOT
By the railroad track between the Devil's Hole and Lewiston.

Senecio aureus, L .... **Golden Ragwort**
Goat Island.
var. **Balsamitæ,** Torr. and Gray.
Near DeVaux College. Rays sometimes wanting.

SENECIO VULGARIS, L .... GROUNDSEL
American side of the River above the Falls.

Erechtites hieracifolia, Raf .... **Fireweed**
Goat Island. Near Clifton, Ontario.

ARCTIUM LAPPA, L... BURDOCK
Goat Island.

CNICUS ARVENSIS, Hoffm........................... CANADA THISTLE
Goat Island and elsewhere.

CNICUS LANCEOLATUS, Hoffm....................... COMMON THISTLE
Goat Island and elsewhere.

Cnicus pumilus, Torr..................... :**Pasture Thistle**
Near DeVaux College.

Cnicus altissimus, Willd... .... **Tall Thistle**
var. **discolor,** Gray.
Goat Island.

Picris echioides, L.
"Along the road side between Clifton and Niagara Falls." *Macoun.*

Lampsana communis, L.
Queenston Heights, Ontario. *Macoun,* on the authority of *Millman.*

Hieracium aurantiacum, L.............................Hawkweed
Goat Island. Introduced.

Hieracium Canadense, Michx..............Canadian Hawkweed
Near DeVaux College.

Hieracium paniculatum, L......................Hawkweed
With the last.

Hieracium venosum, L...........................Hawkweed
Goat Island.

Hieracium scabrum, Michx.....................Hawkweed
Near DeVaux College.

Hieracium Gronovii, L..........................Hawkweed
Goat Island. Near Clifton, Ontario.

Prenanthes alba, L. (*Nabalus albus,* Hook.)........White Lettuce
Goat Island.

Prenanthes altissima, L. (*Nabalus altissimus,* Hook.)
Lewiston. Foster's Flat, Ontario.

Taraxacum officinale, Web................................Dandelion
Goat Island and elsewhere.

Lactuca Canadensis, L....................... **Wild Lettuce**
Goat Island.

Lactuca integrifolia, Bigelow. (*L. Canadensis* L., var. *integrifolia,*
Torr. and Gray.)
Goat Island. Near Clifton, Ontario.

Lactuca leucophæa, Gray. (*Mulgedium leucophœum,* DC.) **Blue
Lettuce.**
Goat Island.

Sonchus oleraceus, L...............................Sow Thistle
Goat Island.

LOBELIACEÆ.

Lobelia syphilitica, L.......................**Blue Lobelia**
On the American side of the river above the Falls.

Lobelia Kalmii, L.......................... **Kalm's Lobelia**
Goat Island. Low grounds near Clifton, Ontario.

Lobelia inflata, L....... .... **Indian Tobacco**
Lewiston and elsewhere.

CAMPANULACEÆ.

Specularia perfoliata, A. DC ...... **Venus's Looking Glass**
Niagara Falls. *Macoun*, on the authority of *Dr. Burgess.*

CAMPANULA RAPUNCULOIDES, L .... BELLWORT
Road sides on the main land. Escaped from cultivation.

Campanula rotundifolia, L .... **Harebell**
Goat Island, and thence to Lewiston.

Campanula aparinoides, Pursh .... **Marsh Bellflower**
Goat Island, in grassy places on the edge of the River.

Campanula Americana, L .... **Tall Bellflower**
Near Clifton, Ontaria.

ERICACEÆ.

Gaylussacia resinosa, Torr. and Gray .... **Huckleberry**
Goat Island, and thence to Lewiston.

Vaccinium stamineum, L .... **Deerberry**
Near DeVaux College and at Lewiston.

Vaccinium vacillans, Solander .... **Low Blueberry**
Goat Island.

Vaccinium corymbosum, L .....,....... **Swamp Blueberry**
Wet ground near Clifton, Ontario.

Arctostaphylos Uva-ursi, Spreng ...'**Bearberry. Killikin-nick.**
Goat Island. Vicinity of the Whirlpool on both sides of the River.

Gaultheria procumbens, L .... **Winter-green**
Near the Whirlpool on the Canadian side.

Chimaphila umbellata, Nutt..' .... **Prince's Pine**
Near DeVaux College.

Pyrola secunda, L.
Near DeVaux College.

Pyrola chlorantha, Swartz.
Niagara Falls. *Clinton.* Near the Whirlpool on the Canadian side.

Pyrola elliptica, Nutt.
Goat Island.

Pyrola rotundifolia, L .... **False Winter-green**
Near DeVaux College.

Pterospora andromedea, Nutt.
Near the Whirlpool on the American side.

Monotropa uniflora, L . . . . Indian Pipe
Near the Whirlpool on the Canadian side.

Monotropa Hypopitys, L . . . . Pine-sap
Goat Island. *Clinton.*

A sphagnous swamp, near Black Creek, Ontario, a few miles south
of Chippewa, has produced the following named plants of this family:

Chiogenes hispidula, Torr. and Gray . . **Creeping Snowberry**

Andromeda polifolia, L . . . . Andromeda

Cassandra calyculata, Don . . . . Leather Leaf.

Ledum latifolium, Ait . . . . Labrador Tea

PRIMULACEÆ.

Dodecatheon Meadia, L. American Cowslip . . . . **Shooting Star**
Goat Island. Introduced.

Steironema ciliatum, Raf.
Goat Island.

Steironema longifolium, Gray.
"Crevices of rocks at Niagara Falls." *Macoun.*

Lysimachia stricta, Ait.
Wet grounds near Clifton, Ontario.

Lysimachia Nummularia, L . . . . Moneywort
Escaping from gardens on the main land.

Lysimachia thyrsiflora.
Wet grounds near Clifton, Ontario.

OLEACEÆ.

Ligustrum vulgare, L . . . . Privet
Well established near Clifton, Ontario.

Syringa vulgaris, L . . . . Lilac
A well-grown lilac-tree was observed in flower in the gorge of the
River, on the Canadian side, near the Cantilever Bridge, where
it could not have been planted by man.

Fraxinus Americana, L . . . . **White Ash**
Goat Island and elsewhere.

Fraxinus sambucifolia, Lam . . . . **Black Ash**
Goat Island. Near Clifton, Ontario.

APOCYNACEÆ.

Apocynum androsæmifolium, L . . . . **Dog Bane**
Near DeVaux College.

Apocynum cannabinum, L.................Indian Hemp
Goat Island. Prospect Park. Lewiston.

VINCA MINOR, L......................PERIWINKLE. CREEPING MYRTLE
Goat Island. Introduced and spreading.

ASCLEPIADACEÆ.

Asclepias tuberosa, L.....................Butterfly Weed
Below the Falls on both sides of the River.

Asclepias incarnata, L.................Swamp Milkweed
Goat Island. Wet grounds near Clifton, Ontario.

Asclepias Cornuti, Decaisne............Common Milkweed
Goat Island and elsewhere.

Asclepias phytolaccoides, Pursh..........Wood Milkweed
Goat Island.

Asclepias quadrifolia, L..........Four-leaved Milkweed
Near DeVaux College.

Asclepias verticillata, L...............Whorled Milkweed
Near DeVaux College.

Acerates viridiflora, Ell........Green-flowered Milkweed
"Niagara Falls." *Macoun.*

GENTIANACEÆ.

Gentiana serrata, Gunner..................Shorn Gentian
Goat Island. Wet grounds near Clifton, Ontario.

Gentiana Andrewsii, Grieseb..............Closed Gentian
Chippewa, Ontario. *Macoun,* on the authority of *Dr. Maclagan.*

Gentiana crinita, Froel, occurs on Islands in Niagara River, near
Lake Erie. *Gentiana Saponaria,* L., may be looked for on either side
of the River above the Falls. *Gentiana quinqueflora,* Lam., probably
will be found on the wooded hillsides near the Whirlpool on the
Canadian side.

Frasera Carolinensis, Walt.............American Columbo
Queenston Heights, Ontario. *Jos. Sturdy. Macoun.*

POLEMONIACEÆ.

Phlox divaricata, L.................Blue Phlox
Goat Island and elsewhere.

Polemonium reptans, L....................Polemonium
Goat Island. Uncommon.

HYDROPHYLLACEÆ.

Hydrophyllum Virginicum, L ....**Water Leaf**
Goat Island. Clifton, Ontario.

Hydrophyllum Canadense, L ....**Water Leaf**
Near Clifton, Ontario.

BORRAGINACEÆ.

CYNOGLOSSUM OFFICINALE, L ....HOUNDS-TONGUE
Goat Island and elsewhere.

Cynoglossum Virginicum, L ....Wild Comfrey
Near DeVaux College.

Echinospermum Virginicum, Lehm. (*Cynoglossum Morisoni,*
DC.)
Goat Island.

ECHINOSPERMUM LAPPULA, Lehm .... STICKWEED
Road sides on the main land.

Myosotis laxa, Lehm ....**Forget-me-not**
Wet grounds, near Clifton, Ontario.

Myosotis verna, Nutt.
Lewiston.

LITHOSPERMUM ARVENSE, L ....CORN GROMWELL
Goat Island and elsewhere.

LITHOSPERMUM OFFICINALE, L .... COMMON GROMWELL
Goat Island and elsewhere.

Lithospermum latifolium, Michx.
Near DeVaux College.

SYMPHYTUM OFFICINALE, L. .... COMFREY
Niagara Falls. *Maroun,* on the authority of *Dr. Maclagan.*

ECHIUM VULGARE, L. ....VIPER'S BUGLOSS
Below the Falls on the American side.

CONVOLVULACEÆ.

IPOMŒA PURPUREA, Lam........ ....MORNING GLORY
Occasionally seen as a garden scape.

CONVOLVULUS ARVENSIS, L.................................BINDWEED.
Above the Falls on the American side. Lewiston.

Convolvulus spithamæus, L. (*Calystegia spithamœa,* Pursh.)
Hedge-Bindweed.
Near the Whirlpool, Ontario. *Clinton.* Abundant near Lewiston.

Convolvulus sepium, L. (*Calystegia sepium,* B. R.) **Wild Morning Glory.**
Below Prospect park. Near DeVaux College.

Cuscuta inflexa, Englemann... .... Dodder
Below Lewiston. Identified by *Dr. Englemann.*

Cuscuta Gronovii, Willd.... .... **Dodder**
Above the Falls on the American side.

Cuscuta Epilinum, Weihe............................ Flax Dodder
Flax fields near Lewiston. *Clinton.*

SOLANACEÆ.

Solanum nigrum, L............................. Common Nightshade
Waste places on the main land.

Solanum Dulcamara, L............................... Bittersweet
Goat Island.

Physalis Virginiana, Mill. (*P. viscosa* L.)....**Ground Cherry**
Below the Falls on the Canadian side.

Lycium vulgare, Dunal.......................... Matrimony Vine
Near the lower Suspension Bridge, on the American side. *Clinton.*

Datura Stramonium, L .... Stramonium. Thorn Apple
Waste places on the main land.

Datura Tatula, L .... Purple Thorn Apple
Near Lewiston.

Hyoscyamus niger, L .... Henbane
Niagara. *Judge Logie.*

SCROPHULARIACEÆ.

Verbascum Thapsus, L .... Mullein
Goat Island and elsewhere.

Verbascum Blattaria, L............................ Moth Mullein
Niagara Falls, Ontario, and between Queenston and Niagara.
Macoun.

Linaria vulgaris, Mill................Butter-and-Eggs.   Toad Flax
Above the Falls on the American side. Lewiston.

Collinsia verna, Nutt.
Goat Island. Introduced.

Scrophularia nodosa, M.......................... **Figwort**
var. **Marilandica,** Gray.
Goat Island.

6

Chelone glabra, L............................Turtle-Head
Wet ground near Clifton, Ontario.

Pentstemon pubescens, Solander...............Pentstemon
Goat Island and elsewhere.

Pentstemon lævigatus, Solander.   (*P. Digitalis*, Nutt.)...Pentstemon.
Goat Island.   Introduced.

Mimulus ringens, L......................Monkey-Flower
Low grounds on the American side above the Falls.

Gratiola Virginiana, L.
With the last.   Gill Creek.   *Clinton.*

Veronica Virginica, L... ....Culver's Physic
Goat Island.   Introduced.

Veronica Anagallis, L...................Water Speedwell
Wet grounds near Clifton, Ontario.

Veronica Americana, Schw.....................Brooklime
Margin of river on the American side above the Falls.

Veronica scutellata, L...................Marsh Speedwell
Wet grounds near Clifton, Ontario.

Veronica officinalis, L..........................Speedwell
Near Lewiston.

Veronica serpyllifolia, L........Thyme-leaved Speedwell
Goat Island.

Veronica peregrina, L.
Above the Falls on the American side.

Veronica arvensis, L ....Corn Speedwell
Goat Island.

Gerardia flava, L.................Downy False Foxglove
Near DeVaux College.

Gerardia quercifolia, Pursh........Smooth False Foxglove
Goat Island.   Near DeVaux College.

Gerardia purpurea, L....................Purple Gerardia
Goat Island.   Wet grounds near Clifton, On'ario.

Gerardia tenuifola, Vahl.
Near DeVaux College.

Castilleia coccinea, Spreng.....................Painted Cup
Near the Whirlpool on the Canadian side.

Pedicularis Canadensis, L......................Lousewort
Goat Island.

Pedicularis lanceolata, Michx.
Wet grounds near Clifton, Ontario.

Melampyrum Americanum, Michx............**Cow-wheat**
Goat Island. Near DeVaux College.

OROBANCHACEÆ.

Aphyllon uniflorum, Gray....................**Broom-rape**
Near the Whirlpool on the Canadian side.

Conopholis Americana, Wallr................**Cancer-root**
Above the Falls on the American side. *Clinton.* "Vicinity of
Niagara Falls." *Macoun*, on the authority of *Dr. Maclagan.*

Epiphegus Virginiana, Bart......................Beechdrops
Goat Island. Not common.

LENTIBULARIACEÆ.

Utricularia vulgaris, L......................**Bladderwort**
Shallow and quiet places in Niagara river, near the American
shore above the Falls.

Utricularia cornuta, Michx.
Wet grounds near Clifton, Ontario. Not common now. "Abundant on the Table Rock." [1818.] *Nuttall, Gen. Am. Pl.,
vol.* 1, *p.* 14.

ACANTHACEÆ.

Dianthera Americana, L..................**Water Willow**
Margin of Niagara River above the Falls on the American side.
Chippewa, Ontario. *Macoun* on the authority of *Dr. Maclagan.*

VERBENACEÆ.

Phryma Leptostachya, L.......................**Lop-seed**
Goat Island.

Verbena urticæfolia, L....................**White Vervain**
Above the Falls on the American side.

Verbena hastata, L.........................**Blue Vervain**
Goat Island and elsewhere.

LABIATÆ.

Teucrium Canadense, L......................**Germander**
Goat Island. Chippewa, Ontario. *Macoun*, on the authority of
Dr. Maclagan.

Ajuga reptans, L...Bugle
Goat Island. Introduced.

Collinsonia Canadensis, L ....Horse Balm
Near Clifton, Ontario.

MENTHA VIRIDIS, L ....SPEARMINT
Goat Island. *Clinton.* Niagara Falls. *Macoun,* on the authority
of *Dr. Burgess.*

MENTHA PIPERITA, L ....PEPPERMINT
Above the Falls on the American side. Near Lewiston.

Mentha Canadensis, L ....Wild Mint
Goat Island and elsewhere.

Lycopus Virginicus, L ....Bugle Weed
Goat Island.

Lycopus sinuatus, Ell. (*L. Europœus,* L., var. *sinuatus,* Gray.)
Shore of Niagara River on the American side above the Falls.

Pycnanthemum lanceolatum, Pursh ....Mountain Mint
Near DeVaux College.

Calamintha Nuttallii, Benth. (*C. glabella,* Benth., var. *Nuttallii,*
Gray.)
In wet, rocky places above Clifton, Ontario.

Calamintha Clinopodium, Benth ....■......Basil
Goat Island.

MELISSA OFFICINALIS, L ....LEMON BALM
Niagara Falls. *Macoun,* on the authority of *Dr. Burgess.*

Hedeoma pulegioides, Pers ....False Penny Royal
Near Lewiston.

SALVIA OFFICINALIS, L ....SAGE
Near DeVaux College. Well established.

Monarda fistulosa, L .... Horse Mint
Goat Island and elsewhere.

Monarda didyma, L., Scarlet Balm, ought to be found in the
low, rich grounds along the River, near Clifton, Ontario.

Lophanthus nepetoides, Benth ....Giant Hyssop
Near DeVaux College. Between Niagara Falls and Lake Ontario.
Macoun, on the authority of *Dr. Maclagan.*

Lophanthus scrophulariæfolius, Benth.
Slopes of Queenston Heights. *Macoun.*

NEPETA CATARIA, L ....CATNEP
Goat Island.

NEPETA GLECHOMA, Benth ....GROUND IVY
Goat Island.

Scutellaria lateriflora, L ...Skull-Cap
Wet grounds near Clifton, Ontario.

Scutellaria parvula, Michx ...Skull-Cap
Near the Whirlpool on the Canadian side.

Scutellaria galericulata, Michx ...Skull-Cap
Goat Island and elsewhere.

Brunella vulgaris, L ...Heal-All
Goat Island and elsewhere.

Physostegia Virginiana, Benth.
Shores of the river above the Falls.

LEONURUS CARDIACA, L ...MOTHERWORT
Goat Island.

Stachys aspera, Michx. (*S. palustris*, L., var. *aspera*, Gray.) **Hedge Nettle.**
Goat Island.

PLANTAGINIACEÆ.

PLANTAGO MAJOR, L ...COMMON PLAINTAIN
Goat Island and elsewhere.

Plantago Rugellii, Decaisne. (*P. Kamtschatica*, Cham.) **Plaintain**
Goat Island and elsewhere.

PLANTAGO LANCEOLATA, L ...RIB GRASS
Near Clifton, Ontario.

PLANTAGO MEDIA, L.
"Niagara." *Provancher, Flore Canadienne, p.* 474. Not seen by us.

AMARANTACEÆ.

AMARANTUS RETROFLEXUS, L ...PIGWEED
Road sides on the main land.

AMARANTHUS ALBUS, L ...WHITE AMARANT
Road sides on the main land.

CHENOPODIACEÆ.

CHENOPODIUM ALBUM, L ...PIGWEED
Goat Island and elsewhere.

CHENOPODIUM GLAUCUM, L.
Road sides and waste places on the main land.

CHENOPODIUM URBICUM, L ...PIGWEED
Bath Island.

CHENOPODIUM HYBRIDUM, L.
Goat Island.

CHENOPODIUM BOTRYS, L........................OAK-OF-JERUSALEM
"Niagara Falls." *Macoun*.

ATRIPLEX PATULA, L......................................ORACHE
var. HASTATA, Gray.
American side of the River above the Falls.
var. LITTORALIS, Gray.
With the last.

PHYTOLACCACEÆ.

Phytolacca decandra, L........................**Poke weed**
Goat Island.

POLYGONACEÆ.

POLYGONUM AVICULARE, L................................KNOT GRASS
Waste places on the main land.

POLYGONUM ERECTUM, L.
⟋ Above the Falls on the American side.

Polygonum incarnatum, Ell.
Shores of the River above the Fall.

Polygonum Pennsylvanicum, L.
"In the Niagara District." *Macoun*, on the authority of *Dr. Maclagan*.

Polygonum amphibium, L.
Islands of Niagara River. *Clinton*.

Polygonum Muhlenbergii, Watson. (*P. amphibium*, L. var. *terrestre*, Willd.)
Margin of the River above the Falls.

POLYGONUM PERSICARIA, L............................LADY'S THUMB
Waste places on the main land.

Polygonum Hydropiper, L..................**Smart weed**
Margin of the River above the Falls.

Polygonum acre, H. B. K....................**Smart weed**
On the American side of the River above the Falls.

Polygonum hydropiperoides, Michx.
Wet places near Clifton, Ontario.

Polygonum arifolium, L....................**Tear-thumb**
Chippewa, Ontario. *Macoun*, on the authority of *Dr. Maclagan*.

Polygonum sagittatum, L....................**Tear-thumb**
Doubtless to be found in the low, wet ground near Clifton, Ontario.

POLYGONUM CONVOLVULUS, L................BLACK BINDWEED
Waste places and road sides of the main land.

Polyonum dumetorum, L..... ....**False Buckwheat**
var. **scandens**, Gray.
Chippewa, Ontario. *Macoun*, on the authority of *Dr. Maclagan.*

RUMEX CRISPUS, L.... .... YELLOW DOCK
Road sides on the main land.

RUMEX OBTUSIFOLIUS, L............................... BITTER DOCK
In similar places as the last.

RUMEX ACETOSELLA, L............................... SHEEP SORREL
Near Clifton, Ontario.

ARISTOLOCHIACEÆ.

Asarum Canadense, L........................**Wild Ginger**
Near Clifton, Ontario. Lewiston. Only the larger form (or species?) noticed.

PIPERACEÆ.

Saururus cernuus, L........................ **Lizard's Tail**
"Charles's Island above the Falls." [Ontario.] *Macoun*, on the authority of *Dr. Burgess.*

LAURACÆ.

Sassafras officinale, Nees........................ **Sassafras**
Lewiston. Near the Whirlpool on the Canadian side.

Lindera Benzoin, Meisner .... **Spice bush**
Goat Island and the Three Sisters.

THYMELACEÆ.

Dirca palustris, L............................ **Moosewood**
Niagara, Ontario. *Macoun*, on the authority of *Dr. Maclagan.*

DAPHNE MEZEREUM, L............................ .... MEZEREON
Goat Island. Introduced and spreading.

ELEAGNACEÆ.

Shepherdia Canadensis, Nutt.......... .... **Shepherdia**
Goat Island and on each side of the River to Lewiston and Queenston.

SANTALACEÆ.

Comandra umbellata, Nutt....:.........**Bastard Toad-Flax**
Goat Island. Lewiston and elsewhere.

EUPHORBIACEÆ.

Euphorbia maculata, L .... **Spotted Spurge**
Main land on both sides of the River.

Euphorbia platyphylla, L .... **Spurge**
Road sides on the main land.

EUPHORBIA HELIOSCOPIA, L .... SPURGE
With the last.

EUPHORBIA CYPARISSIAS, L .... CYPRESS SPURGE
Escaped.

Euphorbia hypericifolia, L .... **Spurge**
Above the Falls on the American side.

Acalypha Virginica, L .... **Three-seeded Mercury**
Near DeVaux College.

CERATOPHYLLACEÆ.

Ceratophyllum demersum, L .... **Hornwort**
In shallow places in the River above the Falls on the American side. In pools near Clifton, Ontario.

URTICACEÆ.

Ulmus fulva, Michx .... **Slippery Elm**
Goat Island. Lewiston.

Ulmus Americana, L .... **American Elm**
Goat Island.

Ulmus racemosa, Thomas .... **Thomas's Elm**
Bath Island. Planted.

ULMUS CAMPESTRIS, L .... ENGLISH ELM,
Luna Island. Planted.

Celtis occidentalis, L .... **Nettle Tree. Sugar Berry**
"Rather common between Queenston and Niagara." *Macoun.*

Morus rubra, L .... **Red Mulberry**
Near DeVaux College. One small specimen observed near the Ferry landing on the Canadian side. "Not uncommon from Niagara Town along the river to the Whirlpool." *Macoun.*

MORUS ALBA, L. . ... WHITE MULBERRY
Spontaneous near Lewiston. ' Niagara Falls. *Macoun.*

MORUS ———
An undetermined species has been planted on Luna Island.

Urtica gracilis, Ait............................Tall Nettle
　Goat Island.

Laportea Canadensis, Gaud...................Wood Nettle
　Damp grounds above Clifton, Ontario.

Pilea pumila, Gray.............. ....Richweed
　Goat Island.

Bœhmeria cylindrica, Willd. has probably been overlooked.

CANNABIS SATIVA, L..HEMP
　Waste places on the main land.

PLATANACEÆ.

Platanus occidentalis, L........Button Wood.　Sycamore
　Goat Island.

JUGLANDACEÆ.

Juglans cinerea, L..............................Butternut
　American side of the River above the Falls.

Juglans nigra, L..........................Black Walnut
　Near DeVaux College. "Niagara Falls." *Macoun.*

Carya alba, Nutt....White Hickory.　Shell-bark Hickory
　Goat Island. Near DeVaux College. "At Queenston Heights
　and the Falls it constitutes the greater part of the forest."
　Macoun.

Carya tomentosa, Nutt......................Hairy Hickory
　"Amongst other hickories in the Niagara peninsula." *Macoun.*

Carya porcina, Nutt.......................Pignut Hickory
　" Queenston Heights and Niagara Falls." *Macoun.*

Carya amara, Nutt........................Bitter Hickory
　Goat Island. In the village near the River. Below Lewiston.

BETULACEÆ.

Betula lenta, L................................Black Birch
　Goat Island.

Betula lutea, Michx...........................Yellow Birch
　Goat Island near the Horse-shoe Fall.

Betula papyracea, Ait.........................Paper Birch
　Goat Island. Below Lewiston.

Alnus incana, Willd...................................Alder
　Goat Island. Wet grounds near Clifton, Ontario.

　7

CUPULIFERÆ.

Carpinus Caroliniana, Walt. (*Carpinus Americana,* Michx.)
Blue Beech.
Goat Island.

Ostrya Virginica, Willd......Iron Wood.   Hop Hornbean
Goat Island.   Some of the trees very large.

Corylus rostrata, Ait ....Hazelnut
Near DeVaux College.

Corylus Avellana, L.......... .. ..European Filbert
Planted on Luna Island.

Fagus ferruginea, Ait................................Beech
Goat Island.   Abundant.

Castanea vulgaris, Lam..........................Chestnut
var. **Americana,** A. DC.
Near DeVaux College.   Lewiston.   Queenston.

Quercus alba, L................................White Oak
Goat Island; but more abundant near DeVaux College.

Quercus obtusiloba, Michx.............. ....Post Oak
" Niagara Falls." *Provancher, Flore Canadienne, p.* 543.

Quercus Prinus, L.........................Chestnut Oak
"Niagara." *Provancher, Flore Canadienne, p.* 543.

Quercus macrocarpa, Michx ....Bur Oak
"Niagara." *Provancher, Flore Canadienne, p.* 543.

Quercus prinoides, Willd.   (*Q. Prinus,* L., var. *humilis,* Gray.)
Dwarf Chestnut Oak.
Goat Island.   "Common on Queenston Heights and in numerous places around Niagara." *Macoun.*

Quercus rubra, L.............................. ...Red Oak
Near Clifton, Ontario.

Quercus coccinea, Wang...................... Scarlet Oak
Goat Island.   "In the forest along the Niagara river it is an abundant tree." *Macoun.*

Quercus tinctoria, Bartram......................Quercitron
Near DeVaux College.   "Not uncommon at Niagara." *Macoun.*

Quercus palustris, Du Roi........................Pin Oak
" Wet woods below Queenston Heights." *Macoun.*

SALICACEÆ.

Salix nigra, Marsh ....Black Willow
Goat Island.   Moist places near Queenston, Ontario. *Macoun.*

Salix lucida, Muhl........................Shining Willow
Goat Island. Near Clifton, Ontario.

Salix discolor, Muhl......................Glaucous Willow
Goat Island.

Salix rostrata, Richardson. (*S. livida,* Wahl., var. *occidentalis,* Gray.) **Livid Willow.**
Above the Falls on both sides of the River.

Salix petiolaris, Smith.
American side of the River above the Falls.

Salix cordata, Muhl..................Heart leaved Willow
Goat Island.

SALIX PURPUREA, L.............. ....Basket Willow
On the American side of the River, above the Falls. "Between Niagara Town and Queenston." *Macoun.*

SALIX ALBA, L., and
SALIX BABYLONICA, L. are not uncommon in cultivation.

POPULUS ALBA, L.............................Abele. White Poplar
Near DeVaux College.

POPULUS CANESCENS, Smith...........................White Poplar
Planted as a shade tree in places. This and the last species have been often confounded by American botanists. *P. canescens* is much the commoner.

Populus tremuloides, Michx..............American Aspen
Goat Island.

Populus grandidentata, Michx.......Large-toothed Aspen
Goat Island. The Three Sisters.

Populus monilifera, Ait......................Cotton wood
Goat Island and occasionally on both sides of the River, above the Falls to Lake Erie.

Populus balsamifera, L.
var. **candicans,** Gray....................Balm of Gilead
Lewiston. A single tree near the Ferry landing on the Canadian side ; probably not planted. The typical form may occur in the vicinity of the Falls as it is not uncommon on the islands in the River near Lake Erie. Not yet observed near the Falls.

POPULUS DILATATA, L.............................Lombardy Poplar
Planted as a shade tree on the main land and spreading by the root.

CONIFERÆ.

Thuja occidentalis, L .... **Arbor-vitæ. White Cedar**
Goat Island. Near DeVaux College. The most abundant of the evergreens growing near the Falls.

Juniperus communis, L .... **Juniper**
Goat Island.

Juniperus Virginiana, L .... **Red Cedar**
Goat Island. Apparently disappearing.

Taxus baccata, L., var. **Canadensis,** Gray .. **Ground Hemlock American Yew.**
Goat Island.

Pinus Strobus, L .... **White Pine**
Goat Island. A few specimens. More plentiful and of larger growth near De Vaux College and below.

Tsuga Canadensis, Carriere. (*Abies Canadensis,* Michx.) . **Hemlock**
Goat Island. Not a prevailing tree.

HYDROCHARIDACEÆ.

Elodea Canadensis, Mich. (*Anacharis Canadensis,* Planch.) ......
Water Snake-Weed.
In the old mill-race above the Falls on the American side and elsewhere.

Vallisneria spiralis, L .... **Tape-grass. Ell-grass**
With the last.

ORCHIDACEÆ.

Corallorhiza multiflora, Nutt .... **Coral root**
Near the Whirlpool, Ontario.

Spiranthes latifolia, Torr .... , ... **Ladies' Tresses**
Wet places near Clifton, Ontario.

Spiranthes cernua, Rich .... **Ladies' Tresses**
In the same places as the last.

Habenaria Hookeri ana, Torr. (*Habenaria Hookeri,* Torr.) **Twa Blade.**
Near the Whirlpool, Ontario.

Habenaria hyperborea, R. Br.
Goat Island near the Horse-shoe Fall.

Cypripedium pubescens, Willd .. **Ladies' Slipper. Moccasin Flower.**
Near De Vaux College.

53

Cypripedium parviflorum, Salisb...**Ladies' Slipper. Moccasin Flower.**
Near the Whirlpool, Ontario.

IRIDACEÆ.

Iris versicolor, L ....**Blue Flag**
Goat Island. In wet places above the Falls on both sides of the River.

Sisyrinchium anceps, Cav. (*Sisyrinchium Bermudiana,* L. var. *anceps,* Gray.)..........................**Blue-eyed Grass**
Near De Vaux College.

Sisyrinchium mucronatum, Mich. (*Sisyrinchium Bermudianv,* L. var. *mucronatum,* Gray.)
Not seen by us. Probably overlooked.

SMILACEÆ.

Smilax herbacea, L......................**Carrion Flower**
Near De Vaux College.

Smilax hispida, Muhl. (**Cat-Brier**), and **Smilax rotundifolia,** L., (**Green Brier**), have not been observed, but probably may be found.

LILIACEÆ.

Allium tricoccum, Ait.......................**Wild Leek**
Goat Island. Abundant.

Allium Canadense, Kalm.
Goat Island. Not common.

Polygonatum biflorum, Ell ....**Solomon's Seal**
Goat Island.

Smilacina racemosa, Desf ....**False Solomon's Seal**
Goat Island and elsewhere.

Smilacina stellata, Desf.
Goat Island. Only the small variety.

Maianthemum Canadense, Desf. (*Smilacina bifolia,* Ker. var. *Canadensis,* Gray.)..................**Two-leaved Solomon's Seal**

ASPARAGUS OFFICINALIS, L ....**ASPARAGUS**
Goat Island and elsewhere. Not common.

Lilium Philadelphicum, L ,.....................**Fire Lily**
Near De Vaux College.

HEMEROCALLIS FULVA, L.................**DAY LILY**
Near the Cantilever bridge on the Canadian side. Escaped from cultivation.

Erythronium Americanum, Ker.. **Yellow Adder's-Tongue**
Goat Island and elsewhere.

Uvularia grandiflora, Smith............................**Bellwort**
Goat Island.

Oakesia sessilifolia, Watson. (*Uvularia sessilifolia,* L.)..... **Small Bellwort.**
Goat Island.

Streptopus roseus, Michx.........................**Streptopus**
Goat Island.

Prosartes lanuginosa, Don........................**Prosartes**
Goat Island.

Veratrum viride Ait...................... ....**False Helebore**
Low grounds near Clifton, Ontario.

Chamælirium Carolinianum, Wild. (*Chamælirium luteum,* Gray.) **Devil's Bit.**
Between Stamford and the Whirlpool, Ontario. Abundant.

Medeola Virginica, L......................**Cucumber Root**
Goat Island. Not common.

Trillium erectum, L......................**Purple Trillium**
Goat Island.
var. **album,** Pursh.
With the typical form.

Trillium grandiflorum, Salisb......**Large White Trillium**
Goat Island. Flower with green stripes through the petals frequently produced.

PONTEDERIACEÆ.

Pontederia cordata, L......................**Pickerel Weed**
Niagara River above the Falls in quiet places.

Schollera graminifolia, Willd............**Water Star Grass**
Not uncommon along the margin of Niagara River near Lake Erie and likely to be found in places nearer the Falls.

JUNCACEÆ.

Luzula pilosa, Willd...................**Hairy Wood-Rush**
Near De Vaux College.

Luzula campestris, DC
With the last.

Juncus effusus, L.............. ....:................**Soft Rush**
Above the Falls near the River on the American side.

Juncus bufonius, L .... **Rush**
In similar situations as the last.

Juncus tenuis, Willd .... **Rush**
Low places on the main land.

Juncus acuminatus, Mich .... **Rush**
var. **debilis,** Engleman.
In low grounds.

Juncus nodosus, L .... **Rush**
Goat Island, in wet places near the margin of the River.
var. **megacephalus,** Torr.
With the last.

Juncus Canadensis, J. Gay .... **Rush**
On both sides of the River above the Falls.

TYPHACEÆ.

Typha latifolia, L .... **Cat-tail Flag**
At the water's edge near the foot of the American staircase.

Typha angustifolia, L .... **Narrow-leaved Cat-tail Flag**
Niagara Falls. *Clinton.*

Sparganium eurycarpum, Engleman .... **Bur Reed**
American side of the River above the Falls.

Sparganium simplex, Hudson .... **Bur Reed**
With the last.
AROIDEÆ.

Arisæma triphyllum, Torr ... **Indian Turnip. Jack-in-the-Pulpit.**
Goat Island. A large and small variety common.

Peltandra Virginica, Raf .... **Arrow Arum**
Niagara Falls. *Clinton.* Not seen by us. Perhaps *P. undulata,* Raf.

Symplocarpus fœtidus, Salisb .... **Skunk's Cabbage**
Wet places above Clifton, Ontario.

Acorus Calamus, L .... **Calamus. Sweet Flag**
With the last.
LEMNACEÆ.

Lemna trisulca, L .... **Duck's meat**
Niagara River above the Falls in quiet places.

Spirodela polyrrhiza, Schleid .... **Duck's meat**
In similar places as the last.

Wolffia Columbiana, Karst.
Niagara River above the Falls. *Prof. Kellicott.*

ALISMACEÆ.

Alisma Plantago, L...................**Water Plantain**
   var. Americana, Gray.
   Margin of Niagara River above the Falls.

Sagittaria variabilis, Engelmann.................Arrow-Head
   Above the Falls in wet places on the Canadian side.

Sagittaria heterophylla, Pursh.
   Niagara River, near La Salle. *Clinton.*

Triglochin palustre, L.
   Wet ground above Clifton, Ontario. Of unusual size.

NAIADACEÆ.

Naias flexilis, Rostk...................**Naiad**
   Niagara River above the Falls.

Zannichellia palustris, L...............**Horned Pondweed**
   Abundant in the upper portion of Niagara River and therefore to
   be expected nearer the Falls.

Potamogeton natans, L.....................**Pondweed**
   Niagara River near Strawberry Island. *Clinton.*

Potamogeton hybridus, Michx...............**Pondweed**
   Black Creek, Ontario, opposite Grand Island.

Potamogeton rufescens, Schrader..............**Pondweed**
   Niagara River. *Rev. Thomas Morong.*

Potamogeton fluitans. Roth. (*P. lonchites*, Tuckerman.)..**Pondweed.**
   Niagara River. *Rev. Thomas Morong.*

Potamogeton lucens, L.....................**Pondweed**
   Niagara River.

Potamogeton amplifolius, Tuckerman...........**Pondweed**
   Niagara River. *Rev. Thomas Morong.*

Potamogeton gramineus, L.................**Pondweed**
   var. **heterophyllus,** Fries.
   Niagara River.
   var. **elongatus,** Morong.
   Niagara River. *Rev. Thomas Morong.*

Potamogeton prælongus, Wulfen..............**Pondweed**
   Niagara. *Provancher, Flore Canadienne,* p. 627.

Potamogeton perfoliatus, L.................**Pondweed**
   var. **lanceolatus,** Robbins.
   Niagara River. *Rev. Thomas Morong.*

57

Potamogeton zosteriæfolius, Schum. (*P. compressus*, L.). Pondweed.
Niagara River.

Potamogeton Niagarensis, Tuckerman .... Pondweed
"Rapids above the Falls." *Gray's Manual, (5th Ed.)* p. 489.
Rediscovered, after many years, in 1886, in the old mill-race
above the Falls, on the American side, by the *Rev. Thomas
Morong.*

Potamogeton pauciflorus, Pursh .... Pondweed
Niagara River.

Potamogeton pusillus, L. .... Pondweed
Niagara River.

Potamogeton pectinatus, L. .... Pondweed
Niagara River.

Potamogeton Robbinsii, Oakes .... Pondweed
Niagara River.

[Note. – It is not asserted that all the species of Potamogeton, above
named, have yet been detected in the immediate vicinity of the Falls, but
as they all occur more or less abundantly in the upper portion of the River
(except as noted), they may well be expected nearer the Falls.]

CYPERACEÆ.

Cyperus diandrus, Torr. .... Galingale
var. castaneus, Torr.
Goat Island, on the east side, near the River.

Cyperus esculentus, L. (*Cyperus phymatodes,* Muhl.).. Galingale
Goat Island near the River.

Cyperus strigosus, L. .... Galingale
Wet places near Clifton, Ontario.

Cyperus filiculmis, Vahl., has probably been overlooked.

Dulichium spathaceum, Richard .... Dulichium
Wet places near Clifton, Ontario.

Eleocharis ovata, R. Br. (*Eleocharis obtusa,* Schult.).. Spike Rush
Near the shores of the River above the Falls.

Eleocharis palustris, R. Br .... Spike Rush
With the last and growing in the water.

Eleocharis tenuis, Schult .... Spike Rush
Damp places above the Falls.

Eleocharis acicularis, R. Br .... Spike Rush
Shores of the River above the Falls on either side.

8

Scirpus planifolius, Muhl......................**Spike Rush**
Near De Vaux College.

Scirpus pungens, Vahl.........................**Spike Rush**
Wet places above Clifton, Ontario.

Scirpus lacustris, L. (*S. validus,* Vahl.)......**Great Bull Rush**
Margin of the River, on the American side, above the Falls. Wet
places above Clifton, Ontario.

Scirpus fluviatilis, Gray.................. ....**Club Rush**
Wet places above Clifton, Ontario.

Scirpus atrovirens, Muhl.
Wet places along the River above the Falls on either side.

Scirpus lineatus, Michx.
East side of Goat Island in wet places near the River.

Eriophorum cyperinum, L, (*Scirpus Eriophorum,* Mich.)..**Wool
Grass.**
Wet places above Clifton, Ontario.

Carex polytrichoides, Muhl.........................**Sedge**
Wet places above Clifton, Ontario.

Carex Steudellii, Kunth............................**Sedge**
Near Clifton, Ontario.

Carex bromoides, Schkr.............................**Sedge**
Wet places near Clifton, Ontario.

Carex vulpinoides, Michx...........................**Sedge**
Above the Falls, on the American side, in low places.

Carex stipata, Muhl................................**Sedge**
With the last.

Carex rosea, Schkr................................**Sedge**
Goat Island in damp places.
var. **retroflexa,** Torr. (*Carex retroflexa,* Muhl.)........**Sedge**
Goat Island in damp places.

Carex sterilis, Willd............................:.....**Sedge**
Near Clifton, Ontario.

Carex scoparia, Schkr..............................**Sedge**
With the last.

Carex tribuloides, Wahl. (*Carex lagopodioides,* Schkr.)...**Sedge**
With the last.
var. **cristata,** Bailey. (*Carex cristata,* Schw.)..........**Sedge**
Goat Island.

Carex straminea, Schkr.. ... .Sedge
Near De Vaux College.

Carex aquatilis, Willd................................Sedge
Margin of the River above the Falls.

Carex torta, Boott.........................:...........Sedge
Near Clifton, Ontario.

Carex stricta, Lam....................................Sedge
In wet, grassy places on the American side above the Falls.

Carex crinita, Lam... ....Sedge
Near Clifton, Ontario.

Carex aurea, Nutt....................................Sedge
Near Clifton, Ontario.

Carex granularis, Muhl.............................Sedge
Near Clifton, Ontario.

Carex conoidea, Schkr...............................Sedge
Moist places on the American side above the Falls.

Carex grisea, Wahl..................................Sedge
With the last.

Carex virescens, Muhl............................... Sedge
Near De Vaux College.

Carex plantaginea, Lam.............................Sedge
Near Lewiston.

Carex retrocurva, Dew... ....Sedge
Goat Island.

Carex platyphylla, Carey...........................Sedge
Near De Vaux College.

Carex digitalis, Willd................ ....Sedge
Goat Island.

Carex laxiflora, Lam............................ ....Sedge
var. plantiginea, Boott.
Near De Vaux College.

Carex eburnea, Boott......... ....Sedge
Goat Island, and near De Vaux College.

Carex Pennsylvanica, Lam.........................Sedge
Lewiston.

Carex varia, Muhl...................................Sedge
Near Clifton, Ontario.

Carex prasina, Wahl. (Carex miliacea, Muhl.)...........Sedge
Wet places above Clifton, Ontario.

Carex debilis, Michx..............................Sedge
With the last.

Carex Œderi, Retz.............................. ....Sedge
Goat Island, near the Horse-shoe Fall. Niagara. *Provancher, Flore Canadienne*, p. 658.

Carex riparia, Curtis..............................Sedge
Eastern side of Goat Island.

Carex trichocarpa, Muhl............................ Sedge
Near Clifton, Ontario.

Carex comosa, Boott................................Sedge
Near Clifton, Ontario.

Carex intumescens, Rudge..........................Sedge
Near Clifton, Ontario.

Carex lupulina, Muhl...............................Sedge
Wet grounds above the Falls on either side.

Carex rostrata, With. var. **utriculata**, Bailey. (*Carex utriculata*, Boott.)...Sedge
With the last.

[NOTE. It is not unlikely that a number of other species of this large and difficult genus may still be found in the vicinity of the Falls; and, as our specimens have not been submitted to any one who has made the *Carices* a special study, it is quite probable that some of our determinations may prove erroneous.]

GRAMINEÆ.

PANICUM GLABRUM, Gaudin............................PANIC GRASS
Road sides on the main land.

PANICUM SANGUINALE, L...................CRAB GRASS. PANIC GRASS
Above the Falls on the American side of the River.

Panicum capillare, L........................**Witch Grass**
A garden weed on the main land.

Panicum virgatum, L.................... ......Panic Grass
Wet grounds near Clifton, Ontario. Dry places near De Vaux College. The latter an unusual situation.

Panicum latifolium, L........................Panic Grass
Near De Vaux College.

Panicum clandestinum, L....................Panic Grass
Goat Island.

Panicum dichotomum, L......................Panic Grass
Near De Vaux College.

Panicum depauperatum, Muhl..... ....Panic Grass
Goat Island and the Three Sisters in rocky places.

Panicum Crus-galli, L.................... ....Barn-Yard Grass
Waste places on the main land.

Setaria viridis, Beauv......................Green Fox-tail Grass
Above the Falls on the American side.

Spartina cynosurioides, Willd..................Cord Grass
Margin of the River above the Falls on either side.

Zizania aquatica, L.............................Wild Rice
Niagara River above the Falls.

Leersia Virginica, Willd.......................White Grass
Goat Island.

Leersia oryzoides, Swartz...................Rice Cut Grass
Wet places above Clifton, Ontario.

Andropogon provincialis, Lam. (*A. furcatus,* Muhl.)....Beard
Grass.
Near De Vaux College.

Andropogon scoparius, Michx.................Beard Grass
Near De Vaux College.

Chrysopogon nutans, Benth. (*Sorghum nutans,* Gray.)..Indian
Grass.
Near DeVaux College.

Phalaris arundinacea, L................Reed Canary Grass
Near Clifton, Ontario. Islands in the River above the Falls.

Oryzopsis melanocarpa, Muhl............. Mountain Rice
Near the Whirlpool, Ontario.

Oryzopsis asperifolia, Michx................Mountain Rice
Goat Island. Near DeVaux College.

Oryzopsis juncea, Michx. (*Oryzopsis Canadensis,* Torr.)..Mountain
Rice.
Near Lewiston.

Muhlenbergia glomerata, Trin.............Drop-seed Grass
Niagara Falls. *Clinton.*

Muhlenbergia Mexicana, Trin.............Drop-seed Grass
Niagara Falls. *Clinton.*

Muhlenbergia sylvatica, Torr. and Gray....Drop-seed Grass
Foster's Flat, Ontario. *Clinton.*

Muhlenbergia Willdenovii, Trin..........Drop-seed Grass
Lewiston.

Muhlenbergia diffusa, Schreb................**Nimble Will**
Near the Whirlpool, Ontaria. *Clinton.*

Muhlenbergia capillaris, Kunth..............**Hair Grass**
"On the talus below the Falls on the American side." *Clinton.*

Brachyelytrum aristatum, Beaud.
Goat Island. Near De Vaux College.

PHLEUM PRATENSE, L.....................................TIMOTHY
Goat Island and the main land.

Alopecurus geniculatus, L ....**Fox-tail Grass**
Near the lower Suspension bridge, Ontario.

Sporobolus vaginæflorus, Vasey. (*Vilfa vaginæflora,* Torr.)..**Rush Grass.**
Lewiston. Near the Whirlpool, Ontario.

Agrostis perennans, Tuckerman.................**Thin Grass**
Lewiston.

Agrostis scabra, Willd.. ....**Hair Grass**
Near DeVaux College.

Agrostis alba, L. (*Agrostis vulgaris,* With.).....**Red Top Grass**
Grassy places. Goat Island and elsewhere.
var. **vulgaris,** Thurb............................. **Fiorin**
Margin of the River above the Falls on the American side.

Cinna arundinacea, L.................. ....**Reed Grass**
Near Clifton, Ontario.

Deyeuxia Canadensis, Beauv. (*Calamagrostis Canadensis,* Beauv.) **Blue-joint Grass.**
Goat Island, on the water's edge.

Deschampsia flexuosa, Griseb. (*Aira flexuosa,* L.)....**Common Hair Grass.**
Near DeVaux College.

Deschampsia cæspitosa, Beauv. (*Aira cæspitosa,* L.)......**Hair Grass.**
Goat Island.

Avena striata, Michx..............................**Wild Oat**
Lewiston and elsewhere.

AVENA SATIVA, L.....................................OAT
Road sides above the Falls. Not persistent.

Danthonia spicata, Beauv..................**Wild Oat Grass**
Near DeVaux College. Near the Whirlpool, Ontario.

Phragmites communis, Trin..........................Reed
Niagara River above the Falls.

Eatonia Pennsylvanica, Gray.....................Eatonia
Moist places above the Falls.

DACTYLIS GLOMERATA, L...........................ORCHARD GRASS
Above the Falls on the American side.

Poa annua, L.............................Low Spear Grass
Goat Island and the main land.

Poa compressa, L..............................Wire Grass
Goat Island and the main land.

Poa serotina, Ehrh........................False Red Top
Wet grounds above Clifton, Ontario.

Poa pratensis, L.....Common Meadow Grass. Kentucky
Blue Grass.
Goat Island.

Poa debilis, Torr.
Near De Vaux College.

Poa alsodes, Gray.
Near De Vaux College.

Glyceria nervata, Trin...............Fowl-Meadow Grass
Above the Falls on both sides of the River.

Glyceria pallida, Trin.
Niagara river above the Falls on the American side.

Glyceria arundinacea, Kunth. (*G. aquatica,* Smith.)......Reed
Meadow Grass.
Wet places above Clifton, Ontario.

Glyceria fluitans, R. Br.
Old mill-race above the Falls on the American side.

Festuca duriuscula, L. (*F. ovina,* L., var. *duriuscula,* Gray.)
Sheep's Fescue.
Goat Island.

FESTUCA ELATIOR, L...............................MEADOW FESCUE
Goat Island.

Festuca nutans, Spreng............................Fescue
Near DeVaux College. Lewiston.

BROMUS SECALINUS, L...............................CHESS
Fields near Lewiston.

Bromus racemosus, L...............................Chess
   Fields near Lewiston.

Bromus Kalmii, Gray.... ....**Wild Chess**
   Near the Whirlpool, Ontario.

Bromus ciliatus, L.......................... ..**Wild Chess**
   Near De Vaux College.

Lolium temulentum, L..........................Bearded Darnel
   Near the landing of the old steamer "Maid of the Mist," on the
   American side. *Clinton.*

Agropyrum repens, Beauv. (*Triticum repens*, L.).......**Couch,
   Quitch, or Quick Grass.**
   Road sides near Clifton, Ontario.

Agropyrum caninum, R. and S. (*Triticum caninum*, L.)..**Awned
   Wheat Grass.**
   Goat Island.

Elymus Virginicus, L........................**Lyme Grass**
   On either side of the River above the Falls.

Elymus Canadensis, L.......................**Lyme Grass**
   Near the Whirlpool, Ontario.

Elymus striatus, Willd ....**Lyme Grass**
   Near De Vaux College.

Asprella Hystrix, Willd. (*Gymostichum Hystrix*, Schreb.)..**Bottle-
   Brush Grass.**

                      EQUISETACEÆ.

Equisetum arvense, L..........................**Horse-Tail**
   Goat Island and elsewhere.

Equisetum limosum, L........................**Horse-Tail**
   Islands of Niagara River. *Clinton.*

Equisetum palustre, L........................**Horse-Tail**
   Wet places above Clifton, Ontario.

Equisetum hyemale, L....................**Scouring Rush**
   Near the Whirlpool, Ontario, and elsewhere.

Equisetum variegatum, Schleicher.
   Goat Island. Wet places above the Falls, on the Canadian side.

Equisetum scirpoides, Michx.
   Near the Whirlpool, Ontario.

OPHIOGLOSSACEÆ.

Botrychium Virginianum, Swartz. (*B. Virginicum*, Swartz.)
Moonwort.
Goat Island and elsewhere.

Botrychium ternatum, Swartz ....Moonwort
var. **intermedium,** D. C. Eaton. (*B. lunarioides*, D. C. Eaton.)
Near Clifton, Ontario.

Ophioglossum vulgatum, L .... Adder's Tongue
Occurs on Grand Island, and may, therefore, be confidently looked
for in favorable situations near the Falls. •

FILICES.

Onoclea Struthiopteris, Hoffm. (*Struthiopteris Germanica*, Willd.)
Ostrich Fern.
Goat Island and elsewhere.

Onoclea sensibilis, L .... Sensitive Fern
Goat Island and elsewhere.

Osmunda regalis, L .... Royal Fern
Goat Island. Not common. Near Clifton, Ontario.

Osmunda Claytoniana, L .... Interrupted Fern
Near Clifton, Ontario.

Osmunda cinnamomea, L .... Cinnamon Fern
Near Clifton, Ontario.

Cystopteris fragilis, Swartz.
Between Lewiston and Youngstown. Near De Vaux College.

Cystopteris bulbifera, Bern .... Bladder Fern
Goat Island and along the American stair-case. Near the Whirlpool, Ontario.

Aspidium Noveboracense, Swartz .... New York Fern
Goat Island.

Aspidium Thelypteris, L ....Shield Fern
Goat Island.

Aspidium spinulosum, Swartz....... .... Shield Fern
var. **intermedium,** D. C. Eaton.
Near Clifton, Ontario.

Aspidium cristatum, Swartz.................. ....Shield Fern
Foster's Flat, Ontario.

9

Aspidium Goldianum, Hook...................**Shield Fern**
Near the Whirlpool, Ontario.

Aspidium marginale, Swartz...................**Shield Fern**
Goat Island.   Devil's Hole.   Foster's Flat, Ontario.

Aspidium Lonchitis, Swartz...................**Shield Fern**
"Sparingly at Foster's Flat," Ontario. *Burgess.*

Aspidium achrostichoides, Swartz.........**Christmas Fern**
Goat Island.   Near De Vaux College.   Lewiston.

Phegopteris Dryopteris, Fée...................**Beech Fern**
Devil's Hole.

Camptosorus rhizophyllus, Link...........**Walking Fern**
Not uncommon near the Whirlpool, Ontario.   Foster's Flat,
Ontario.   Near De Vaux College, but rare.

Scolopendrium vulgare, Smith.........**Hart's tongue Fern**
Introduced in one place in the gorge of the River by Judge
Clinton several years ago.

Asplenium Trichomanes, L...................**Spleen wort**
Near De Vaux College.   Foster's Flat, Ontario.

Asplenium ebeneum, Ait... ....**Spleen wort**
Near De Vaux College, but rare.   Near Lewiston, abundant.

Asplenium achrostichoides, Swartz. (*Asplenium thelypteroides,*
Michx.)..**Spleen wort**
Near the Whirlpool, Ontario.

Asplenium Filix-fœmina, Bern.................**Lady-Fern**
Foster's Flat, Ontario.

Pellæa gracilis, Hook..........................**Cliff-Brake**
"Crevices of rocks at Foster's Flat," Ontario. *Burgess.* Not
seen by us.

Pellæa atropurpurea, Link.....................**Cliff-Brake**
Formerly on Goat Island and the Three Sisters.   Not lately seen
by us.   Probably extirpated.   Near De Vaux College.   Foster's
Flat, Ontario.   Rare.

Pteris aquilina, L...................... .....**Common Brake**
Near De Vaux College and at Lewiston.

Adiantum pedatum, L.................**Maiden-Hair Fern**
Goat Island.   Not abundant.   Near Clifton, Ontario.

Polypodium vulgare, L............................**Polypody**
Goat Island.   The Three Sisters.   Near De Vaux College.
Lewiston.

www.ingramcontent.com/pod-product-compliance
Lightning Source LLC
Chambersburg PA
CBHW022105210326
41519CB00056B/1432